Western Technology and China's Industrial Development

Hsien-ch'un Wang

Western Technology and China's Industrial Development

Steamship Building in Nineteenth-Century China, 1828–1895

Hsien-ch'un Wang
Institute of History
National Tsing Hua University
Hsinchu, Taiwan

ISBN 978-1-137-59903-2   ISBN 978-1-137-59813-4 (eBook)
https://doi.org/10.1057/978-1-137-59813-4

© The Editor(s) (if applicable) and The Author(s) 2022
This work is subject to copyright. All rights are solely and exclusively licensed by the Publisher, whether the whole or part of the material is concerned, specifically the rights of translation, reprinting, reuse of illustrations, recitation, broadcasting, reproduction on microfilms or in any other physical way, and transmission or information storage and retrieval, electronic adaptation, computer software, or by similar or dissimilar methodology now known or hereafter developed.
The use of general descriptive names, registered names, trademarks, service marks, etc. in this publication does not imply, even in the absence of a specific statement, that such names are exempt from the relevant protective laws and regulations and therefore free for general use.
The publisher, the authors and the editors are safe to assume that the advice and information in this book are believed to be true and accurate at the date of publication. Neither the publisher nor the authors or the editors give a warranty, expressed or implied, with respect to the material contained herein or for any errors or omissions that may have been made. The publisher remains neutral with regard to jurisdictional claims in published maps and institutional affiliations.

This Palgrave Macmillan imprint is published by the registered company Springer Nature America, Inc.
The registered company address is: 1 New York Plaza, New York, NY 10004, U.S.A.

To my mother, my brother, and the memory of my father

ACKNOWLEDGEMENTS

There are so many people to thank for help and encouragement that made this book possible. I owe so much to Professor David Faure for his intellectual stimulation and guidance. He never ceases to be a source of inspiration. Tao Tao Liu, Laura Dewby, Rajeswary Ampalavanar Brown, Rana Mitter, and David Pong were extremely helpful throughout my years at Oxford. Yung-fa Chen, Yi-Long Huang, Hsiang-lin Lei, Jen-der Lee, Man-houng Lin, and especially Susan Naquin were most encouraging and helpful at the crucial point of my career. I would like to thank Chi-Kong Lai and Catherine Jami who invited me to contribute conference papers from which I developed parts of this book. I am also grateful to Alexei Volkov, Yi-Long Huang, Daiwie Fu, and Hsiao-pei Yen for reading the drafts and giving me precious suggestions. There was a time when I actually abandoned the project, Min-ching Tseng and Hsiang-ke Chao urged me to resume it. I would also like to thank Hsin-yi Lin, Hsiao-ting Lin, Vera Dorofeeva-Lichtmann, as well as Sarah and Steven Snowden, who are always close and supportive.

Librarians and archivists of the University of Oxford's Bodleian Library, Cambridge University Library, the Public Record Office in London, the Institute for the History of Natural Sciences Library in Beijing, Shanghai Library, Academia Sinica's Kuo Ting-yee Library and Fu Ssu-nien Library, the National Palace Museum Library in Taipei, and National Tsing Hua University Library in Hsinchu were most helpful. I must thank the Wai Seng Scholarship of St. Antony's College, Oxford and Monsoon Asia Project of National Tsing Hua University, Taiwan for funding my trips to the archives and libraries in the UK, China, and Taiwan. The earlier version of Chap. 2 has been published in *Technology and Culture*. Map 5.1 in Chap. 5 has been published in David Pong's *Shen Pao-chen and China's Modernisation in the Nineteenth Century*. Figure 2.1 in Chap. 2 is published in Joseph Needham's *Science and Civilisation in China*, volume 4, part 2. Figure 5.1 in Chap. 5 is published in Richard Wright's *The Chinese Steam Navy*. I am grateful to John Hopkins University Press, Cambridge University Press, and Greenhill Books for permission to republish them in this book.

My debt to my family is tremendous. My father never entertained the idea that I would become a historian. After his untimely death, my mother supported my studies although she suffered from bad health. I am extremely grateful to my brother for taking up the responsibility of taking care of my mother, allowing me to spend much of my time on my work. To them I dedicate this book.

Abstract

This book explores one of the most discussed but poorly understood topics of modern Chinese history, the transfer of Western technology into China in the nineteenth century. For decades historians of modern China often contend that, due to political or cultural conservatism, Qing China failed to learn Western technology in the second half of the nineteenth century. In recent years Benjamin Elman has challenged that argument by showing the Chinese elite's efforts to absorb Western mathematics and natural philosophy since the sixteenth century. Nevertheless, the fundamental question of how Western technology was transferred into China is yet to be answered.

Steamship technology was a result of centuries-long techno-scientific development in Western Europe. The steam engine, whose valved piston-cylinder mechanism turned the pressure differential within and without the cylinder into a motive force. Its production required machine tools in precision metal works; it benefited from Europe's tradition of using technical drawing as a means of visual communication. Wood was replaced by iron, and later steel, as the major shipbuilding material because of its strength and sustainability. Furthermore, in a neck-breaking speed of technological progress, thermodynamics played a crucial role.

Chinese scholar-officials were interested in observing nature and studying mathematics. Yet, their understanding of nature employed a completely different set of concepts and terms. Although they might have been involved in managing engineering works, they were not engineers.

Besides, Chinese artisans did not excel in metal cutting. They heavily relied on the oral tradition in passing on knowledge; even if they drew to communicate, they used very different drawing conventions from what their European counterparts employed. How did Qing dynasty China learn the skills and knowledge of steamship building?

Chapter 1 introduces the research problem, emphasizing the scientific and technological underpinnings of the steamship building, comparing them with Chinese technical traditions. Chapter 2 traces the history of China's exposure to Western steamships during the Opium Wars (1839–1842) and the impact of this exposure. Early attempts were made to build steamers; these efforts did not succeed in building them but taught China to appreciate the importance of machine tools in their production. By the mid-1860s, when the Qing government decided to recruit European technicians to build steamships, senior officials had understood the importance of steam power in the role of strengthening the empire's military strength.

Chapter 3 explores how the nineteenth-century concept of heat was introduced into China. From the 1850s, foreign missionaries and their Chinese collaborators had started to translate popularized Western scientific and technological texts into Chinese. The texts introduced new ideas to tackle old conceptions about heat. Chapter 4 examines how the translations of steam-engine management textbooks and popularized science texts had introduced more new terms and concepts to the Chinese readership. Yet, these texts confused force (*li*) with energy because the translators' failure to find a terminological solution to translate energy. Furthermore, because theses texts were not used in school education, their impact on Chinese understanding of the science of heat was limited.

Chapter 5 examines the role of the Fuzhou Navy Yard, established in 1866 by the Qing government, in training skilled workers and engineers. It shows that the navy yard recruited foreign technicians and teachers to teach technical drawing, mechanical engineering, and mathematics to Chinese apprentices and students who had not previously been exposed to Western science and technology. Some students were sent for training in Europe; and they became China's first generation of marine engineers. By 1895, the naval yard's Chinese technicians were able to build steamers without foreign supervision. Chapter 6 examines how the shift of government policy made an impact on domestic shipbuilding programme. After the Sino-French War in 1884, Qing officials realized that war demanded the most advanced technology and hence opted to purchase warships from

Western Europe. The cost of the procurement strategy, however, was to divert precious funds away from much-needed upgrading in the Fuzhou Naval Yard. Therefore, it may be argued that, without an explicit policy, the Qing government had decided that it would be satisfied to leave its shipyards behind international standards.

Chapter 7 concludes that, by 1895, China had made technical progresses in steamship building. Yet, because the Qing government's institutional reforms in education and industrial funding were not radical enough to allow the most advanced scientific knowledge to be introduced and more capital-intensive shipbuilding enterprises to be initiated, China's shipbuilding industry was limited to only a couple of government shipyards and the technical knowledge was only passed on to a small number of engineers and workers. Radical reforms only came in 1904 when the Qing government abolished the imperial examination and introduced the company law.

Keywords   Ding Gongchen • *Huangbu* • Xu Shou • Hua Hengfang • Li Honghang • Zeng Guofan • Jiangnan Arsenal • Benjamin Hobson • W.A.P. Martin • John Fryer • Alexander Wylie • Joseph Edkins • Young John Allen • *Tongwen guan* • *Gezhi shuyuan* • *Gezhi keyi* • Prosper Giquel • Fuzhou Navy Yard • Fuzhou Naval School • *Beiyang* Fleet

Contents

1 Introduction — 1

2 Discovering Steam Power in China, 1828–1865 — 23

3 Translating Heat: Tackling Old Conceptions with New Ideas, 1855–1868 — 57

4 Achievements and Constraints of Late Qing Translations of Heat, 1868–1895 — 95

5 Training Workers and Engineers: The Fuzhou Navy Yard, 1866–1895 — 135

6 To Buy Or to Build? — 175

7 Conclusion — 209

Glossary — 219

Index — **225**

List of Figures

Fig. 2.1	Chang Qing's wheel boat. (Source: Joseph Needham, *Science and Civilisation in China,* volume 4, part 2, (Cambridge University Press, 1965), figure 638)	27
Fig. 2.2	Ding Gongchen's steam cylinder. (Source: Ding Gongchen, *Yapao tushuo jiyao* (1945 reprint of the 1843–1851 edition, kept in the Xiamen University Library))	31
Fig. 2.3	Ding Gonchen's steam locomotive. (Source: Ding Gongchen, *Yapao tushuo jiyao* (1945 reprint of the 1843–1851 edition, stored in the Xiamen University Lirbary))	32
Fig. 2.4	Zheng Fuguang's steamboat diagram. (Source: Zheng Fuguang, *Jingjing lingchi* 3:37b (Taipei: Yiwen yinshuguan, 1966, reprint of the nineteenth century edition))	35
Fig. 3.1	The *Bowu xinbian*'s illustrations of the thermometers. Source: He Xin, *Bowu xinbian* (undated edition kept in the Kuo Ting-yee Library, Academia Sinica, Taipei)	63
Fig. 3.2	A page from the *Xia'er guanzhen*. Source: Sheng Guowei, Uchida Keiichi, Matsuura Akira (eds.), *Xia'er guanzhen, fu jieti, suoyin* (Shanghai: Cishu chubanshe, 2005), p. 496(223)	69
Fig. 4.1	The illustration of the calorimeter in the *Rexue tushuo*. (Source: Fu Lanya (John Fryer), *Rexue tushuo* 2:1a, (the 1890 edition kept in the Shanghai Library))	110

Fig. 4.2   The illustration of the calorimeter provided by the essayist Yang Yuhui in the 1890 Chinese Prize Essay Contest. (Source: *Gezhi shuyuan keyi*, Volume 2 (Shanghai: Shanghai kexue jishu wenxian chubanshe, 2016), p. 373.) 115

Fig. 5.1   The *Pingyuan* after it was captured by the Japanese Imperial Navy. Source: Richard N.J. Wright, *The Chinese Steam Navy, 1862–1945* (London: Chatham Publishing, 2000), p. 81 159

CHAPTER 1

Introduction

In 1842, British guns and ships defeated the Qing dynasty of China in what has since been known as the Opium War (1839–1942). Given the superiority of Western armament, one might have presumed that the experience would have prompted the Qing government to acquire the new technology which played a significant role in its defeat. Yet, the Qing government only did so after another Opium War (1856–1860) and the Taiping Rebellion (1851–1865) in the 1860s. Historians blame that delay on the Qing government being in its decline, inept in facing up to the technological challenge.[1] That argument culminated in John King Fairbank's Western stimulus-China response model and dominated the Modern Chinese History field before it gave way to the skepticism bred by the distrust of "Orientalism". The contemporary China historian is more likely to seek an explanation for Qing China's woeful experiences in the hands of foreign powers in the internal socio-economic or political dynamics. Nevertheless, elements of political ideology, sometimes referred to as Confucianism, education, politics, and economy have not yielded an explanation either for Qing China's technological backwardness or for its tardiness in taking up the new technology coming from the West. This book is driven by an urge to break that impasse. While, like many China historians, I do not accept that the technological imbalance of the mid-nineteenth century between China and the West necessarily reflected backwardness that may be generalized to many other aspects of Chinese culture, I begin from the position that that imbalance is undeniable. China historians, much like some of the Qing dynasty officials they criticize, do not seem to appreciate the fact that the history of technology has to be studied in relation to its techno-scientific settings.

THE STEAM ENGINE AND ITS TECHNO-SCIENTIFIC BACKGROUND

The technology that China needed to import was the steam engine, the prime mover of the Industrial Revolution from the late eighteenth century. Historians of Chinese science and technology have often argued, for example in Joseph Needham's "Pre-Natal History of the Steam-Engine", that elements of the steam engine, including the concept of void, atmospheric pressure, and the piston-cylinder mechanism, had already existed in China and, somehow made their way to Western Europe.[2] This argument suggests a linear evolution from the humble valved cylinder-piston mechanism, either be the ancient Greek cast-bronze water pump or the traditional Chinese square wooden furnace bellows, to the steam engine invented by Thomas Newcomen (1664–1729). Such an argument fails to consider the unique techno-scientific environment in which the Newcomen engine was invented. The key is not just the use of the cylinder-piston mechanism but also its role in natural philosophers' quest for understanding the phenomena of vacuum and the atmosphere pressure from the seventeenth century. As Graham Hollister-Short, H. Floris Cohen, and others have shown, Newcomen's invention was an embodiment of that techno-scientific culture in which the cylinder-piston mechanism was deliberately used to utilize the difference between the atmospheric and steam pressure for the purpose of lifting a weight.[3]

Furthermore, efficiency lied in the core of engine building. Because the Newcomen engine was prone to breakdown and was extremely inefficient, James Watt (1736–1819)'s addition of an external condenser to it required the machine tools tradition that materialized the idea. By the mid-nineteenth century, thanks to the power of the steam engine and the rising demand for more sophisticated machinery, more machine tools of different descriptions were invented, such as the steam hammer, which could use the power of steam to forge large metal parts, the high precision screw cutting lathe, which could reach the accuracy of a million of an inch, and the milling machine that produced complex shapes of metal parts. The advancement in machine tools made interchangeable parts possible, and that, in turn, gave birth to modern mass production. As L.T.C. Rolt, Robert Woodbury, and Nathan Rosenberg have pointed out, the significance of machine tools in engineering lay in their capacity for precision, whether working on large or delicate metal machine parts.[4]

To meet the ever-rising demand for iron in machine construction, metallurgy progressed with the assistance of the steam engine. R.F. Tylecote has shown that by the early nineteenth century, high-quality wrought iron produced by the reverberatory furnace (for re-smelting pig iron produced by the blast furnace) was responsible for the strength of British firearms and engineering. Furthermore, in the nineteenth century, the steam engine helped to increase iron output, rolling out iron sheets in a scale and homogeneity that was unattainable by the water wheel or human hands. Besides, after the 1850s, thanks to the introduction of much more economical method of steel making, steel gradually replaced iron as the major material for machine construction.[5]

Technical drawings became the crucial tool of production control by the nineteenth century when machines became increasingly sophisticated. As David McGee, John Brown and Ken Alder have shown, drawings were effective tool of visual communication in conveying the sense of shapes, colours, and space that words are unable to do.[6] Renaissance architects had started to use drawings to communicate with masons. Shipwrights also employed drawings in the process of designing ships. In the eighteenth century, geometricians invented descriptive geometry, giving the drawing practices not only a theoretical basis but also mathematical precision. Since then, technical drawings based on the principle of parallel projection were introduced in arsenals and factories and were standardized and employed by machine designers to transmit detailed instructions to workmen producing the machinery.

Furthermore, one of the important features of nineteenth century development of technology was the application of mathematicized scientific theories to the production process. As D. S. L. Cardwell and Crosbie Smith have shown, the engineer's paramount concern to increase the efficiency of the steam engine spurred their interest in the science of thermodynamics. Energy, pressure, temperature, and mechanical work came to be related by the application of calculus in the design of efficient engines. From a time when machines might have been put together by skilled craftsmen who had some scientific knowledge to the period when the building of engine required the application of scientific knowledge and mathematical techniques of infinitesimal calculus, Western Europe went through no less than a technological revolution which had implications not only for social differentiation but, more specifically, in the education of the engineer.[7]

In the aftermath of the First Opium War, some Chinese literati had noticed that Britain's victory might be captured in the term *chuanjian paoli* (sturdy ships and penetrating guns). Yet, to learn to produce Western-style guns and build steamships, they had to acquire an entire technology with all its ramifications, from its theoretical bases to practical application, from the machine tools that cut metal parts with good precision, the drawings that bridged production and design, to the scientific theory that made the design of an efficient engine possible. This was a technology that was very far from China's own.

Historians have noted that from the sixteenth century onwards, the Ming and later Qing courts valued the astronomical and mathematical knowledge brought in by the missionaries of the Society of Jesus.[8] Chinese scholar-officials worked with the Jesuits, translating texts such as Euclid's *Elements* into Chinese. Yet, the cultural exchange between Europe and China was largely cut off in the eighteenth century over the controversy of whether traditional Chinese rites were qualified as religious ones.[9] In addition, the science of energy that was invented in the mid-nineteenth century was beyond anything the Jesuits and their Chinese colleagues could have grasped.

Furthermore, although Chinese artisans could reach a high standard in metal casting, they did not excel in metal cutting.[10] Chinese artisans used the lathe to cut wood, jade, or ivory but not metal. Besides, although the Jesuit missionaries might have introduced the techniques and the tools of clock making into China, Chinese clocksmiths were few in number and the technology was not diffused to other handicrafts.[11] How might China transferred steam engine technology from that technological settings?

SELF-STRENGTHENING

Until the eighteenth century, the Qing dynasty government was interested in employing Western practical knowledge, especially firearm technology.[12] It remains a mystery how in the first half of the nineteenth century, it was caught practically unaware of the sea change in Western technology. Despite the makeshift defence put up by Imperial Commissioner Lin Zexu in Guangzhou, the Qing government was unprepared for either British fire-power or mobility in the First Opium War, and it took another crisis, incurred concurrently by the Taiping Rebellion and the Second Opium War to force the Qing government to reconsider its technological backwardness. When, towards the end of the 1850s, the protracted rebellion threatened the very existence of Qing rule effect, officials who led the defense started to equip their troops with foreign firearms. By 1860, Beijing was overrun by British and French troops, forcing the court of the Xianfeng Emperor (r. 1850–1861) into exile. In 1861, Zeng Guofan, a senior official who led the anti-Taiping campaign, suggested to the throne that China would gain from "learning the barbarians' wisdom in producing guns and building ships."[13] Prince Gong, the Xianfeng Emperor's half-brother who was left behind in Beijing by the fleeing court to negotiate the peace settlement with the invaders, memorialized to the throne for the institution of new ways of managing foreign affairs, which included training students in foreign languages and instituting a new department to deal with foreign affairs (to be known as the *Zongli yamen* or the Office in Charge of Affairs Concerning All Nations).[14] The term "self-strengthening" (*ziqiang*) was introduced in this context.[15] The emperor gave permission to both proposals.

Zeng Guofan and his lieutenant Li Hongzhang established arsenals in 1861 and 1863 respectively to provide their troops with the necessary ammunitions. Yet, they lacked proper machinery and technicians. In 1864 Li Hongzhang, the most ardent advocate for introducing Western military technology, proposed to the *Zongli yamen* that "[i]f China wants to strengthen itself (*ziqiang*), the best way is to learn [how to make] foreign weaponry."[16] The *Zongli yamen* agreed: "[w]e find that the way to govern the state lies in self-strengthening. ... The most important matter regarding self-strengthening is to train troops, but the manufacture of weaponry must precede the training of troops."[17] Therefore, the three aspects of the self-strengthening policy were: training diplomatic and translation talents to obtain foreign knowledge, training an army, and supplying it with Western-style firearms and steamers. A direct result of this policy was the establishment of the Jiangnan Arsenal (*Jiangnan zhizao ju*) by Li Hongzhang in 1866 and the Fuzhou Navy Yard (*Fuzhou chuanzheng ju*) by Zuo Zongtang, another distinguished anti-Taiping official, in 1867. Both institutions were large-scale industrial undertakings far exceeding earlier efforts. They hired foreign technicians and imported machine tools from the West for manufacturing firearms and building steamships. More arsenals were established over the next twenty years or so.

Qing officials also established educational institutions for training translators and diplomats. In 1861, Prince Gong set up the School of Combined Learning (*Tongwen guan*) and added the courses of mathematics and astronomy to its curriculum in 1867. In 1863 Li Hongzhang established the School for the Diffusion of Languages (*Guang fangyan guan*) in Shanghai and its curriculum included mathematics and physics. In 1869, the school was incorporated into the Jiangnan Arsenal.[18] In 1868, Zeng Guofan founded the Translation Department (*Fanyig uan*) for translating Western works on a wide range of subjects, including military skills, politics, and technology, into Chinese.[19]

The 1860s was an extraordinary time in the Qing dynasty's history. The Xianfeng Emperor had died in 1861, and the child Tongzhi Emperor (r. 1862–1875) reigned under the influence of the Empress Dowager Cixi rather than his regents. Cixi's politics required her to ally with the forward-looking Prince Gong, who worked closely with senior officials such as Zeng Guofan, Zuo Zongtang, and Li Hongzhang. At heart, the Taiping Rebellion had destroyed the Manchu banners and the Green Standard as creditable military forces and seriously damaged government finance. The rebellions had been crushed by militia forces led by provincial officials such as Zeng, Zuo and Li, and supported by local taxes such as internal transit dues (*lijin*).[20] The imperial court had retained the right of appointment, but its financial and military weakness were paving the way for the rise of regionalism, a term marked by the financial and administrative independence of provincial governor-generals and governors in the second half of the nineteenth century even as they remained loyal to the imperial cause.

Although the imperial court was well aware that the country's security depended on the success of the self-strengthening program, it had little direct knowledge of Western science and technology, and, therefore, relied on its forward-looking senior officials for advice. It was also preoccupied with rebellions: the Taipings until 1865, the Nians until 1868, and the Moslem uprising in the northwest until 1877. Nevertheless, however forward-looking, those senior officials were themselves only in the process of realizing that science and technology underpinned Western military strength. To think that they could, at this stage of their knowledge, have opted for a complete overhaul of education and government to cater to the foreign knowledge would be an anachronistic reading of contemporary history.

Yet, as early as in September 1865, Li Hongzhang, in memorializing the court on the establishment of the Jiangnan Arsenal, had argued that the steam engine could be used in agriculture, fabric weaving and spinning, printing, etc.[21] This deviation from the narrowly focused military self-strengthening policy soon appeared in official debate by 1866 to 1867, when senior officials argued, in response to suggestions made by Western diplomats posted in Beijing, over whether the railway and the telegraph should be built. They acknowledged the benefits of the technologies of the railway and the telegraph but were cautious about allowing foreigners to build them.[22] There was no overall policy of "self-strengthening" as such, but through such debates, the terms of reference for "self-strengthening" had broadened.

The broadening of the "self-strengthening" discourse was helped by spread of Western know-how in the treaty ports. Through the 1860s, the publication of books by missionaries and newspapers such as *The North China Herald* and the *Shenbao* had provided both commoners and officials with information about foreign countries. Besides, with the increase of overseas trade, foreign-run steam navigation companies, machine shops, and dockyards had been established in the treaty ports.[23] Many of them, especially the steam navigation companies, were in the form of joint-stock companies incorporated under Western law, issuing shares for pooling capital.[24] They were examples to Chinese merchants and officials of capital-intensive Western enterprises employing the new machinery and new business organization. It was soon realized that enterprises on the Western model could be profitable. As private enterprises using Western technology had not explicitly been permitted by the Qing government, those senior officials who could obtain imperial permission to launch enterprises making use of the new technology brought under their wings Chinese entrepreneurs who were sensitive to the business potentials of Western technology. The result was the combination of share capital and official patronage known as "official sponsorship and merchant management" (*guandu shangban*). Under this formula, in 1872, Li Hongzhang and his protégés, mostly merchants with foreign trade background and purchased degrees, formed the China Merchants' Steam Navigation Company with imperial approval. This was the first Chinese business of scale organized in the form of a joint-stock company, which after initial problems, became highly profitable to its share-holders from the 1880s.[25] Li Hongzhang promoted other business enterprises, starting from 1879, with a telegraphic line near his official seat Tianjin, Zhili province. The telegraphic network grew, and was managed under the Telegraph Bureau set up under Li's aegis in 1880.

Nevertheless, even as senior officials ventured into business, national defense remained the major political issue. In 1874, Japan invaded the island of Taiwan. Coupled with the war against Moslem rebellion in the northwest which was still raging, the crisis triggered a debate in 1875 on whether the government should put more resources into frontier or maritime defense. The imperial court came to the conclusion that maritime defense was as important as frontier defense, and a Maritime Defense Fund was formed to support warship purchasing. However, the fund was insufficient to coordinate the separate naval programs. Following developments from the 1870s, the imperial court decreed a naval policy of resting maritime defense on two squadrons, the Northern Sea (*Beiyang*) Fleet and the Southern Sea (*Nanyang*) Fleet. They were both to purchase warships from abroad. However, as John Rawlinson has shown, there was little coordination between the two fleets. Because of the lack of funds and the strategic importance of the *Beiyang* fleet in defending the capital, it had priority in receiving funds.[26]

In the political climate of the 1870s and 1880s, therefore, when defense considerations mingled with business opportunities, senior officials talked about wealth and power (*fuqiang*) being advanced by Western technology.[27] A memorial by Li Hongzhang in 1875, elaborated on how Western technology could strengthen the country's military strength as it increased its wealth. In the case of private enterprise in shipping, he reiterated an argument that linked business closely to defense:

> Since [China] opened trade with Western countries, [the number of foreign] steamships and sailing ships increases everyday. They travel very fast, hence the profits of inland and maritime shipping are almost monopolized by foreigners. Besides, maritime defence cannot be progressively deployed without steamships. [Therefore, we] must encourage the people to obtain steamships. At peaceful times [the steamships] can ship tribute rice, passengers and commodities; when there is war they can transport relief troops and ammunition. By doing so the difficulties of merchants and the people can be relieved and the morale of self-strengthening will be boosted.[28]

Li argued that since China Merchants' Steam Navigation Company was established, the company's steamships had managed to claimed "right to profit" (*liquan*) from foreigners. The company's fleet also participated in military operations during the Taiwan Crisis in 1874, transporting troops and supplies. Therefore, by linking the two issues, Li Hongzhang successfully presented a case for a business operation under the cloak of self-strengthening. Other industries followed. Industries such as mining could be directly linked to the military agenda, because, especially after the Taiwan Crisis, senior officials were looking for domestic alternatives to imported iron and coal to feed the growing demand for these materials from arsenals, naval fleets, and the China Merchants' Steam Navigation Company.[29] When the Kaiping Coal Mine was established in 1881 by Li Hongzhang and his associates, it supplied coal to the China Merchants' Steam Navigation Company and China's modern naval forces. Yet, it was also an "official-sponsorship-and-merchant-management" (*guandu shangban*) enterprise substantially financed by share capital. By 1890, there were three major modern enterprises established under Li Hongzhang's patronage.[30]

From the mid-1870s, other senior officials took part in the "self-strengthening" enterprises. A notable example of such was Zhang Zhidong, governor-general of Guangdong and Guangxi from 1884 to 1889. When, in 1889, the prospect for building railways in China opened up, he memorialized the court proposing an iron mine in Guangdong. Linking business profit to the national agenda of self-strengthening, he wrote, "[m]y humble opinion is that today's method of self-strengthening is to open up the source of profit and to prevent profit from flowing out."[31] Nevertheless, he was less successful than Li Hongzhang in building power base out of the new enterprises. He ordered the necessary machinery from Britain, including a Bessemer converter, to build an ironworks with a steel production arm in Guangdong. However, soon after the imperial court acceded to his proposal of building the Bejing-Hankou Railway, he was also transferred to the governor-generalship of Hunan and Hubei. In anticipation of the approval, and recognizing that railway building would consume a substantial amount of iron and steel, he moved all the steel making machinery to Hanyang, Hubei province, to form the Hanyang Ironworks. It used the iron ore from Daye, Hubei. However, the Bessemer converter was ill-suited to the Daye iron ore, which contained a high percentage of phosphorus. In consequence, the Hanyang steel was too brittle to be used neither in railway building nor gun making. Hence, the ironworks remained a money-losing undertaking, forcing the Qing government to suspend its running in 1894. Qing officials only completed re-equipping it in 1907 with funds raised through obtaining loans and issuing shares.[32]

It should come as no surprise, therefore, that while senior officials remained loyal to the throne, they also manipulated the politics of self-strengthening to advance their own pet projects and political influence. The imperial court and the ministries were barely able to keep up with the new enterprises, let alone co-ordinate their programs in weapon procurement, technology import, or manpower training. Just as Ting-yee Kuo and Kwang-ching Liu have pointed out, "Self-strengthening became less a rallying cry for genuine efforts at innovation than a shibboleth that served to justify expenditures and vested bureaucratic interests."[33] At best, self-strengthening embodied only a vague ideology; and at its worst, it was made up only of a language acceptable to both the imperial court and provincial authorities.

Transferring Western Technology into China

The successful transfer of technology depended less on political rhetoric than on the movement and training of technicians and workmen, the indigenisation of new skills and knowledge and the provision of funds on the scale required for new enterprises. Scott Montgomery reminds us that the cross-cultural transmission of scientific knowledge from ancient Greek masters, Roman authors, and medieval Arabic thinkers, to Renaissance European natural philosophers required translators to compile and reinterpret earlier wisdoms and transmit them in their own languages.[34] David Landes has stressed the importance of the movement of technicians was essential Continental Europe's emulating the Industrial Revolution originated in Britain.[35] Nathan Rosenberg, Robert Fox, Anna Guagnini and others have also noted that a well-established education system is essential in technology transfer and industrial development.[36] Furthermore, Phyllis Deane, François Crouzet, Rondo Cameron, and others have shown that institutions that pooled capital for industrial investment was one of the factors that financed large-scale projects from cotton spinning-weaving factories to railway in Britain, Germany, Russian and Japan.[37]

Japan's experience in the nineteenth century corroborates these views. By the time the US steam warships forced Japan to open for trade in 1853, Japanese scholars had been keen learners of both Chinese culture and Western learning through translation.[38] Yet, nineteenth-century science and technology required much more than translation and scholarly exchanges. Foreign technicians provided essential help to Japan's industrialisation after the Meiji Restoration (1868).[39] Although the precise role of the Meiji government in Japan's industrialization is still a matter of debate, historians largely agree that it was instrumental in launching many large large-scale projects such as shipyards, arsenals, railways, and the telegraph.[40] More important, As John Bartholomew, Richard Rubinger and others have pointed out, the Meiji government's educational reforms were essential to Japan's scientific and technological progress.[41]

In China, similar government-backed enterprises made good technical progress, as Thomas Kennedy, David Pong, and Marianne Bastid have shown.[42] Even after the Sino-Japanese War of 1895, production continued at the arsenals and shipyards; the steam navigation company carried on its business; the coal mines, ironworks, and the telegraph bureau were operating, and railways continued to be built. They survived into the Republican era and became the backbone of China's heavy industry.[43]

Besides, as James Reardon-Anderson, Wang Yangzong, David Wright, Michael Lackner, Iwo Amelung, and many others have documented, the nineteenth-century translation of Western scientific texts into Chinese brought new terms and new ideas, especially in the fields of chemistry and physics, into Chinese lexicon.[44] More recently, Feng Shanshan has given a very detailed account of the introduction of thermal science into Chinese through translation.[45]

Nevertheless, China historians have not yet answered how technology was transplanted to China from its theoretical principles to its practical applications. This book seeks to document that process by examining how steam engine technology was transferred into China through developing a steamship industry. Chapter 2 explores how the Chinese people discovered the steam engine as a source of power. Chapter 3 discusses how the basic concepts of nineteenth-century theories of heat were translated into Chinese in the 1850s and 60s. Foreign and Chinese translators worked together, negotiating the difference between the linguistic and conceptual differences between ninetieth-century theories of heat and Chinese natural philosophy. Chapter 4 explores the achievements and constraints of late Qing translations of thermal science before 1895. Although the Chinese literati were interested in reading translated texts on the steam engine and thermal science, their understanding of it and its impact was restricted by limited educational reform. Chapter 5 discusses how China institutionalized steam technology by examining the case of the Fuzhou Navy Yard, established in 1866. Chapter 7 discusses the constraint imposed by the shortage of capital in the promotion of a domestic shipbuilding industry. Caught in an arms race with Japan, the Qing government found it imperative to purchase its warships from the West rather than building them in its own shipyards. The diversion of funds from shipbuilding dented the technological acquisition effort.

In other words, this book seeks to demonstrate that the history of technological transfer at the end of the Qing dynasty can be understood in relation to the state of knowledge of Western science and technology held by senior officials, their politics, and the availability of capital. What it does not do is to characterize that effort as a "movement" the results of which may be discussed in terms of "success" or "failure."[46] Nor does it indulge in culturalism as a catch-all explanation of a poorly understood process. Ian Inkster has pointed out that such arguments in the history of technology transfer are redundant and wrong.[47] Just as Benjamin Elman has demonstrated the benefits to understanding the late Qing by placing the self-strengthening efforts within China's scientific tradition, this book attempts to do so in relation to scientific and technological changes.[48] There may or may not be substance to the cultural argument in the backdrop, but until we understand how the science and technology had changed, we are not in a position to put that argument to the test.

Notes

1. Prominent examples are Ssu-yü Teng and John King Fairbank, *China's Response to the West: A Documentary Survey, 1839–1923* (Cambridge, MA: Harvard University Press, 1954); and John Rawlinson, *China's Struggle for Naval Development* (Cambridge, MA: Harvard University Press, 1967).
2. Joseph Needham, "Pre-Natal History of the Steam-Engine," *Transactions of the Newcomen Society*, 35:1 (1962), pp. 3–58.
3. For the history of the cylinder-piston pump, see Graham Hollister-Short, "On the Origins of the Suction Lift Pump," *History of Technology* 15 (1993), pp. 57–75; Graham Hollister-Short, "The Formation of Knowledge Concerning Atmospheric Pressure and Steam Power in Europe from Aleotti (1589) to Papin (1690)," *History of Technology* 25 (2004), pp. 137–150; M. T. Wright, "On the Lift Pump," *History of Technology* 18 (1996), pp. 13–37; H. Floris Cohen, "Inside Newcomen's Fire Engine, or: The Scientific Revolution and the Rise of the Modern World," *History of Technology* 25 (2004), pp. 111–132.
4. L.T.C. Rolt, *Tools for the Job: A History of Machine Tools to 1950* (London: HMSO Publications Centre, 1986); Nathan Rosenberg, "Technological Change in the Machine Tool Industry," *The Journal of Economic History*

23:4 (1963), pp. 414–443; Robert S. Woodbury, *Studies in the History of Machine Tools* (Cambridge, MA: The MIT Press, 1973). The idea of interchangeable parts was invented by the French general Jean-Baptiste de Gribeauval, who intended to standardize weapons with standardized parts in 1765. In 1785, Thomas Jefferson, then the United States Minister to France, witnessed the French gunsmith Honoré Blanc assembling a musket from uniformly produced parts and saw the potential of interchangeability in arms production. He brought the idea to the United States. However, the realization of interchangeability was only realized by the increasingly high precision of the late nineteenth century machine tools. See Ken Alder, *Engineering Revolution: Arms and Enlightenment in France 1663–1815* (Stanford: Stanford University Press, 1997); David A. Hounshell, *From the American System to Mass Production, 1800–1932: The Development of Manufacturing Technology in the United States* (Baltimore, Maryland, USA: Johns Hopkins University Press, 1984).
5. R.F. Tylecote, *A History of Metallurgy* (London: Institute of Materials, 1992); R. Chadwick, "The Working of Metals," in Charles Singer et al. (eds.) *A History of Technology* vol. 5 The Late Nineteenth Century, c1850–1900 (Oxford: Oxford University Press, 1958), pp. 605–635.
6. An overview of the history can be found in Ken Baynes and Francis Pugh, *The Art of the Engineer* (Guildford, England, 1981). For the role of technical drawings in manufacturing, see David McGee, "From Craftsmanship to Draftsmanship: Naval Architecture and the Three Traditions of Early Modern Design," *Technology and Culture* 40 (1999), pp. 209–236; Ken Alder, *Engineering Revolution: Arms, and Enlightenment in France, 1763–1815* (Princeton: Princeton University Press, 1997), pp. 136–143. For the development of technical drawing techniques, see Peter Jeffrey Booker, *A History of Engineer Drawing* (London: Northgate Publishing Co., 1979).
7. D.S.L. Cardwell, *From Watt to Clausius: The Rise of Thermodynamics in the Early Industrial Age* (Ithaca, N.Y.: Cornell University Press, 1971); Crosbie Smith, *The Science of Energy: A Cultural History of Energy Physics in Victorian Britain* (Chicago: Chicago University Press, 1998).
8. Catherin Jami, *The Emperor's New Mathematics: Western Learning and Imperial Authority during the Kangxi Reign (1662–1722)* (Oxford: Oxford University Press, 2012); Benjamin A. Elman, *On Their Own Terms: Science in China 1550–1900* (Cambridge, MA: Harvard University Press, 2005).
9. Paul Rule, "Towards a History of the Chinese Rites Controversy," in David E. Mungello (ed.), *The Chinese Rites Controversy: Its History and Meaning* (Nettetal: Steyler Verlag, 1994), pp. 249–266.
10. Hua Jueming, *Zhongguo gudai jinshu jishu: tong he tie zaojiu de wenming* (Zhengzhou: Daxiang chubanshe, 1999).
11. I am grateful to Drs. Wang Cheng-hua and Lai Hui-min of Academia Sinica and Ms. Chang Li-tuan of the National Palace Museum, Taipei, who point out to me that artisans of the Qing court could have used the lathe

to produce delicate ball-shaped objects. Unfortunately, the tools are still nowhere to be seen. It is possible that the Jesuits and the European artisans who served in the imperial court introduced machine tools to imperial workshop artisans. Catherine Pagani has documented the Jesuits and European clocksmiths had produced clocks and watches for the Qing court and officials. However, she does not discuss the clock making tools. Chinese artisans do not seem to have further developed the principle of the lathe in producing larger metal machine parts. Catherine Pagani, *Eastern Magnificence and European Ingenuity: Clocks of Late Imperial China* (Ann Arbor, 2001).

12. Joanna Waley-Cohen, "China and Western Technology in the Late Eighteenth Century," *American Historical Review* 98:5 (1993), pp. 1525–1544.
13. *Chouban yiwu shimo*, Xianfeng reign, 71:12a.
14. *Chouban yiwu shimo*, Xianfeng reign, 71:17ff; 72:9b–10b; 72:11a. The complete title of the *Zongli yamen* was *Zongli geguo tongshang shiwu yamen* (the Office for General Management of Trade and Affairs with All Countries).
15. The phrase "self-strengthening" is taken from the *Book of Change* (*Yi jing*): "The movement of heaven is full of power. Thus, the gentleman strengthens himself ceaselessly." Kwang-ching Liu, "The Beginnings of China's Modernization," in Samuel C. Chu and Kwang-Ching Liu (eds.), *Li Hung-chang and China's Early Modernization* (Armonk: M.E. Sharpe, 1994), pp. 5–6.
16. *Chouban yiwu shimo*, Tongzhi reign, 25:10b.
17. *Chouban yiwu shimo*, Tongzhi reign, 23:1a.
18. Knight Biggerstaff, *The Earliest Modern Government Schools in China* (Ithaca: Cornell University Press, 1961) and Xiong Yuezhi, *Xixue dongjian yu wanqing shehui* (Shanghai: Shanghai renmin, 1994).
19. Xiong Yuezhi, *Xixue dongjian yu wanqing shehui*, pp. 493–550; Benjamin Elman, *On Their Own Terms: Science in China, 1550–1900*, pp. 360–368.
20. For the rise of regionalism, see Stanley Spector, *Li Hung-chang and the Huai Army: A Study in Nineteenth-Century Chinese Regionalism* (Seattle: University of Washington Press, 1964) and Franz Michael, "State and Society in Nineteenth Century China," *World Politics* 7:3 (1955), pp. 419–433.
21. Zhongguo shixuehui (ed.), *Yangwu yundong*, vol. 4 (Shanghai: Shanghai renmin chubanshe, 1961), p. 14.
22. Although some senior officials rejected the idea of railway building per se, Li Hognzhang and Shen Baozhen, then the director of the Fuzhou Navy yard, considered that China must adopt railways. *Chouban yiwu shimo*, Tongzhi reign, 53:5a; *Chouban yiwu shimo*, Tongzhi reign, 55:13a–14a.

23. Throughout the 1870s, the visibility of the newspaper's advertisements for various machinery, especially that for the cotton spinning and weaving industry, was high. The number of similar advertisements significantly decreased in the 1880s but the discussion of industry surfaced again in the early 1890s.
24. Kwang-ching Liu, "Steamship Enterprise in Nineteenth-Century China," *Journal of Asian Studies* 18:4 (1959), pp. 438–439; Wang Jingyu, "Shijiu shiji waiguo qinhua qiye zhong de huashang fugu huodong," in Huang Yiping, *Zhongguo jindai jingjishi lunwen xuan* (Shanghai: Shanghai renmin, 1985), p. 197.
25. For more details, see Chi-kong Lai, "Li Hung-chang and Modern Enterprise: The China Merchants' Company, 1872–1885," in Samuel C. Chu & Kwang-ching Liu (eds), *Li Hung-chang and China's Early Modernization*, pp. 216–247.
26. John Rawlinson, *China's Struggle for Naval Development, 1839–1895* (Cambridge, MA: Harvard University Press, 1967).
27. Li Hongzhang, *Li Wenzhong gong quanji*, memorials, 24:1ff.
28. It was common that Chinese merchants became shareholders in foreign-run shipping companies. Li Hongzhang considered that such behaviours did not conform to government policy. Li Honghzang, *Li Wenzhong gong quanji*, Memorial, 25:4a–5a.
29. Sun Yutang (ed.), *Zhongguo jindai gongyeshi ziliao diyi ji 1840–1895* (Taipei: Wenhai, 1979, reprint of the 1957 edition), p. 567.
30. Apart from the China Merchants' Steam Navigation Company and the Kaiping Mines, the Shanghai Cotton Cloth Company was established through the same process. The promotion of the company started in 1878 but went into production in 1890. Zhang Guohui, *Yangwu yundong yu jindai qiye* (Beijing: Zhongguo shehui kexue chubanshe, 1979), pp. 272–279.
31. *Yangwu yundong*, vol. 7, p. 203.
32. In 1896, in an attempt to avert the fate of the undertaking, the government's put it under the management of Sheng Xuanhui, one of most successful Chinese entrepreneurs in the late nineteenth and the early twentieth century. However, Sheng and his foreign technicians only discovered the chemical composition of the Daye iron ore in 1904 and then started to reequip ironworks. In 1908, Shen merged the ironworks with the Daye Iron Ore and the Pingxiang Coal Mine, Hubei that provided coals to the ironworks and formed the *Hanyeping* Coal and Iron Company. For detailed discussions, see Albert Feuerwerker, "China's Nineteenth Century Industrialization: the Case of the Hanyeping Coal and Iron Company, Limited," in C. D. Cowan (ed.), *The Economic Development of China and Japan: Studies in Economic History and Political Economy* (New York:

Praeger, 1964), pp. 79–110; Quan Hansheng, *Hanyeping gongsi shilue* (Hong Kong: Chinese University of Hong Kong, 1972).
33. Kwang-ching Liu and Ting-yee Kuo, "Self-strengthening: the Pursuit of Western Technology," in *Cambridge History of China* vol. 10 (Cambridge: Cambridge University Press, 1978), pp. 491–542.
34. Scott L. Montgomery, *Science in Translation: Movements of Knowledge through Cultures and Time* (Chicago: The University of Chicago Press, 2000).
35. David S. Landes, *The Unbound Prometheus: Technological Change and Industrial Development in Western Europe from 1750 to the Present* (Cambridge: Cambridge University Press, 1969), pp. 139–142, 147–150.
36. Nathan Rosenberg, "The International Transfer of Technology: Implications for the Industrialized Countries," in Nathan Rosenberg, *Inside the Black Box: Technology and Economics* (Cambridge: Cambridge University Press, 1982), pp. 245–279; Robert Fox and Anna Guagnini (eds.), *Education, Technology and Industrial Performance in Europe, 1850–1939* (Cambridge: Cambridge University Press, 1993).
37. Phyllis Deane, "The Role of Capital in the Industrial Revolution," *Exploration in Economic History*, 10 (1972), pp. 349–364; Francois Crouzet, "Capital Formation in Great Britain during the Industrial Revolution," in *idem, Britain Ascendant: Comparative Studies in Franco-British Economic History* (Cambridge: Cambridge University Press, 1990), pp. 149–212; Rondo Cameron (ed.), *Banking in the Early Stages of Industrialization* (New York: Oxford University Press, 1967); Kozo Yamamura, "Japan 1868–1930: A Revised View," in Steen Tolliday, *The Economic Development of Modern Japan, 1868–1945: From the Meiji Restoration to the Second World War* (Cheltenham: Elgar, 2001) vol. 2, pp. 168–187. More detailed discussion can be found in Rondo Cameron (ed.), *Financing Industrialization*, 2 vols. (Aldershot: Edward Elgar, 1992).
38. Masayoshi Sugimoto and David L. Swain, *Science and Culture in Traditional Japan* (Rutland, Vermont: Charles E. Tuttle Company, 1989); Grant K. Goodman, *Japan and the Dutch, 1600–1853* (Richmond, Surrey: Curzon, 2000); A. Querido, "Dutch Transfer of Knowledge Through Deshima: The Role of the Dutch in Japan's Scientific and Technological Development During the Edo Period," *Transactions of the Asiatic Society of Japan*, 3rd series, 18 (1983), pp. 17–37.
39. Hazel Jones, *Live Machine: Hired Foreigners and Meiji Japan* (Vancouver: University of British Columbia Press, 1980). For the role of the education system in Japan's technological development, see Toshio Toyoda (ed.) *Vocational Education in the Industrialization of Japan* (Tokyo, United Nations University Press, 1987).

40. Thomas C. Smith, *Political Change and Industrial Development in Japan: Government Enterprise, 1868–1880* (Stanford: Stanford University Press, 1968, reprint of 1955 edition); Tetsuro Nakaoka, "The European Industrial Economy and the Endogenous Development in Asia", in Yamada Keiji (ed.), *The Transfer of Science and Technology Between Europe and Asia* (Kyoto & Osaka: International Research Center for Japanese Studies, 1994); Kozo Yamamura, "Success Illgotten? The Role of Militarism in Japan's Technological Progress," *The Journal of Economic History* 31:1 (1977), pp. 113–135.
41. John R. Bartholomew, *Formation of Science in Japan: Building A Research Tradition* (Yale: Yale University Press, 1989); Richard Rubinger, "Education: From One Room to One System," in Marius B. Jansen and Gilbert Rozman (eds.), *Japan in Transition: From Tokugawa to Meiji* (Princeton: Princeton University Press, 1988), pp. 195–230; Tessa Morris-Suzuki, *The Technological Transformation of Japan: From the Seventeenth to the Twenty-first Century* (New York: Cambridge University Press, 1994), pp. 71–157; Toshio Shishido, "Japanese Industrial Development and Policies for Science and Technology," *Science*, New Series 219:4582 (1983), pp. 259–264; G.C. Allen, "Education, Science, and Economic Development of Japan," *Oxford Review of Education* 4:1 (1978), pp. 27–36.
42. Thomas L. Kennedy, *The Arms of Kiangnan: Modernization in the Chinese Ordnance Industry, 1860–1895* (Boulder, Colorado: Westview Press, 1978); David Pong, *Shen Pao-chen and China's Modernization in the Nineteenth Century* (Cambridge: Cambridge University Press, 1994); Basidi (Marianne Bastid), "Fuzhou chuanzheng ju de jishu yinjin," *Chuanshi yanjiu* No.10 (1996), pp. 104–114.
43. Li Peide, "Lun Jiangnan zhizaoju juwu fenjia de jingyingshi yiyi," *Jindaishi xuekan* 2 (2015), pp. 1–9; Jiangnan zhaochuanchangshi bianxiezhu, *Jiangnan zhaochuanchang shi* (Shanghai: Shanghai renmin chubanshe, 1975).
44. Zhang Zigao and Yang Gen, "Cong *Huaxue chujie* he *Huaxue jianyuan* kan Zhongguo zaoqi fanyi de huaxue shuji he huaxue minci," in Yang Gen (ed.), *Xu Shou he Zhongguo jindai huaxue shi* (Beijing: Kexue jishu wenxian chubanshe, 1986), pp. 105–118; Zhang Hao, "Zai chuantong yu chuangxin zhijian: shijiu shiji de zhongwen huaxue yuansu minci," *Huaxue jiaoyu* 59:1 (2001), pp. 51–59; Wang Yangzong, "A New Inquiry into the Translation of Chemical Terms by John Fryer and Xu Shou," in Michael Lackner, Iwo Amelung, and Joachim Kurtz (eds.) *New Terms for New Idea: Western Knowledge and Lexical Change in Late Imperial China* (Leiden: Brill, 2001), pp. 271–283.

45. Feng Shanshan, "Jindai xifang rexue zai Zhongguo de chuanbo, 1855–1902,", Ph.D. thesis, Inner Mongolia Normal University (2019).
46. The word "movement" (*yundong*) as used in the phrase, "Self-strengthening Movement" (*Ziqiang yundong*), was introduced in the 1930s, most prominently by Jiang Tingfu. In his *Modern Chinese History* (*Zhongguo jindai shi*, Changsha, Hunan: Wenshi yanjiu hui, 1938), Jiang devoted a chapter to "Self-strengthening and its failure," to present self-strengthening as modernization (*jindai hua*). He ascribed the failure of the movement to conservatism. Jiang eloquently presented the Self-strengthening Movement and its framework as a historical fact. See also Wang Xinzhong, "Fuzhou chuanchang zhi yange," *The Tsing Hua Journal* 8:1 (1932), p. 6. and Jiang Tingfu, "Zhongguo yu jindai shijie de da bianju," *The Tsing Hua Journal* 9:4 (1934), pp. 825–826.
47. Ian Inkster, "Prometheus Bound: Technology and Industrialization in Japan, China and India Prior to 1914 – A Political Economy Approach," *Annals of Science* 45 (1988), p. 400.
48. Benjamin Elman, *On Their Own Terms: Science in China, 1500–1900*.

REFERENCES

Alder, Ken, *Engineering Revolution: Arms and Enlightenment in France 1663–1815* (Stanford: Stanford University Press, 1997).

Allen, G.C., "Education, Science, and Economic Development of Japan," *Oxford Review of Education* 4:1 (1978), pp. 27–36.

Baojun 寶鋆 et al. (comp.), *Chouban yiwu shimo* 籌辦夷務始末, Tongzhi reign 同治朝 (Beipeing: National Palace Museum, 1930, reprint of 1880 edition).

Bartholomew, John R., *Formation of Science in Japan: Building A Research Tradition* (Yale: Yale University Press, 1989).

Basidi 巴斯蒂(Marianne Bastid), "Fuzhou chuanzheng ju de jishu yinjin 福州船政局的技术引进," *Chuanshi yanjiu* 船史研究 No.10 (1996), pp. 104–114.

Baynes, Ken, and Pugh, Francis, *The Art of the Engineer* (Guildford, England, 1981).

Biggerstaff, Knight, *The Earliest Modern Government Schools in China* (Ithaca, N.Y.: Cornell University Press, 1961).

Booker, Peter Jeffrey, *A History of Engineering Drawing* (London: Northgate Publishing Co., 1979).

Cameron, Rondo (ed.), *Banking in the Early Stages of Industrialization* (New York: Oxford University Press, 1967).

Cameron, Rondo (ed.), *Financing Industrialization*, 2 vols. (Aldershot: Edward Elgar, 1992).

Cardwell, D.S.L., *From Watt to Clausius: The Rise of Thermodynamics in the Early Industrial Age* (Ithaca, NY: Cornell University Press, 1971).

Chadwick, R., "The Working of Metals," in Charles Singer et al. (eds.) *A History of Technology* vol. 5 The Late Nineteenth Century, c1850–1900 (Oxford: Oxford University Press, 1958), pp. 605–635.

Crouzet, François, "Capital Formation in Great Britain during the Industrial Revolution," in *idem, Britain Ascendant: Comparative Studies in Franco-British Economic History* (Cambridge: Cambridge University Press, 1990), pp. 149–212.

Deane, Phyllis, "The Role of Capital in the Industrial Revolution," *Exploration in Economic History*, 10 (1972), pp. 349–364.

Elman, Benjamin, *On Their Own Terms: Science in China, 1550–1900* (Cambridge, Mass: Harvard University Press, 2005).

Feng Shanshan 冯珊珊, "Jindai xifang rexue zai Zhongguo de chuanbo, 1855–1902 近代西方热学在中国的传播, 1855–1920,", PhD thesis, Inner Mongolia Normal University (2019).

Feuerwerker, Albert, "China's Nineteenth Century Industrialization: the Case of the Hanyeping Coal and Iron Company, Limited," in C. D. Cowan (ed.), *The Economic Development of China and Japan: Studies in Economic History and Political Economy* (New York: Praeger, 1964), pp. 79–110.

Floris, Cohen H., "Inside Newcomen's Fire Engine, or: The Scientific Revolution and the Rise of the Modern World," *History of Technology* 25 (2004), pp. 111–132.

Fox, Robert and Guagnini, Anna (eds.), *Education, Technology and Industrial Performance in Europe, 1850–1939* (Cambridge: Cambridge University Press, 1993).

Goodman, Grant K., *Japan and the Dutch, 1600–1853* (Richmond, Surrey: Curzon, 2000).

Hollister-Short, Graham, "On the Origins of the Suction Lift Pump," *History of Technology* 15 (1993), pp. 57–75.

Hollister-Short, Graham, "The Formation of Knowledge Concerning Atmospheric Pressure and Steam Power in Europe from Aleotti (1589) to Papin (1690)," *History of Technology* 25 (2004), pp. 137–150.

Hounshell, David A, *From the American System to Mass Production, 1800–1932: The Development of Manufacturing Technology in the United States* (Baltimore, Maryland, USA: Johns Hopkins University Press, 1984).

Hua Jueming 华觉明, *Zhongguo gudai jinshu jishu: tong he tie zaojiu de wenming* 中国古代金属技术:铜和铁的文明 (Zhengzhou: Daxiang chubanshe, 1999).

Inkster, Ian, "Prometheus Bound: Technology and Industrialization in Japan, China and India Prior to 1914 – A Political Economy Approach," *Annals of Science* 45 (1988), pp. 399–426.

Jami, Catherin, *The Emperor's New Mathematics: Western Learning and Imperial Authority during the Kangxi Reign (1662–1722)* (Oxford: Oxford University Press, 2012).

Jiang Tingfu 蔣廷黻, "Zhongguo yu jindai shijie de da bianju 中國與近代世界的大變局," *The Tsing Hua Journal* 清華學報 9:4 (1934), pp. 825–826.

Jiang Tingfu 蔣廷黻, *Zhongguo jindai shi* 中國近代史 (Changsha, Hunan: Wenshi yanjiuhui, 1938).

Jiangnan zhaochuanchang shi bianxiezhu 江南造船厂史编写组, *Jiangnan zhaochuanchang shi* 江南造船厂史(Shanghai: Shanghai renmin chubanshe, 1975).

Jiazhen 賈楨 et al. (compl.), *Chouban yiwu shimo* 籌辦夷務始末, Xianfeng reign 咸豐朝 (Beiping: the Palace Museum, 1930, reprint of 1867 edition).

Jones, Hazel, *Live Machine: Hired Foreigners and Meiji Japan* (Vancouver: University of British Columbia Press, 1980).

Kennedy, Thomas L., *The Arms of Kiangnan: Modernization in the Chinese Ordnance Industry, 1860–1895* (Boulder, Colorado: Westview Press, 1978).

Lai, Chi-kong, "Li Hung-chang and Modern Enterprise: The China Merchants' Company, 1872–1885," in Samuel C Chu & Kwang-ching Liu (eds), *Li Hung-chang and China's Early Modernization* (Armonk, New York: M. E. Sharpe, 1994), pp. 216–247.

Landes, David, *The Unbound Prometheus: Technological Change and Industrial Development in Western Europe from 1750 to the Present* (Cambridge: Cambridge University Press, 1969).

Li Hongzhang 李鴻章, *Li Wenzhong gong quanji* 李文忠公全集 (Jinlin: 1908).

Li Peide 李培德, "Lun Jiangnan zhizaoju juwu fenjia de jingyingshi yiyi 论江南制造局局坞分家的经营史," *Jindaishi xuekan* 2 (2015), pp. 1–9.

Liu, Kwang-ching and Kuo, Ting-yee, "Self-strengthening: the Pursuit of Western Technology," *Cambridge History of China*, vol.10 pt.1 (Cambridge: Cambridge University Press, 1978), pp. 491–542.

Liu, Kwang-ching, "Steamship Enterprise in Nineteenth-Century China," *Journal of Asian Studies*, 18:4 (1959), pp. 446–459.

Liu, Kwang-ching, "The Beginnings of China's Modernization," in Samuel C. Chu and Kwang-ching Liu (eds.), *Li Hung-chang and China's Early Modernization* (Armonk: M.E. Sharpe, 1994), pp. 3–14.

McGee, David, "From Craftsmanship to Draftsmanship: Naval Architecture and the Three Traditions of Early Modern Design," *Technology and Culture* 40 (1999) pp. 209–236.

Michael, Franz, "State and Society in Nineteenth Century China," *World Politics*, 7:3 (1955), pp. 419–433.

Montgomery, Scott L., *Science in Translation: Movements of Knowledge through Cultures and Time* (Chicago: The University of Chicago Press, 2000).

Morris-Suzuki, Tessa, *The Technological Transformation of Japan: From the Seventeenth to the Twenty-first Century* (New York: Cambridge University Press, 1994).

Nakaoka, Tetsuro, "The European Industrial Economy and the Endogenous Development in Asia," in Yamada Keiji (ed.), *The Transfer of Science and*

Technology Between Europe and Asia (Kyoto & Osaka: International Research Center for Japanese Studies, 1994).

Needham, Joseph, "Pre-Natal History of the Steam-Engine," *Transactions of the Newcomen Society*, 35:1 (1962), pp. 3–58.

Pagani, Catherine, *Eastern Magnificence and European Ingenuity: Clocks of Late Imperial China* (Ann Arbor: University of Michigan Press, 2001).

Pong, David, *Shen Pao-chen and China's Modernization in the Nineteenth Century* (Cambridge: Cambridge University Press, 1994).

Quan Hansheng 全漢昇, *Hanyeping gongsi shilue* 漢冶萍公司史略 (Hong Kong: Chinese University of Hong Kong, 1972).

Querido, A., "Dutch Transfer of Knowledge Through Deshima: The Role of the Dutch in Japan's Scientific and Technological Development During the Edo Period," *Transactions of the Asiatic Society of Japan*, 3rd series, 18 (1983), pp. 17–37.

Rawlinson, John L., *China's Struggle for Naval Development, 1839–1895* (Cambridge, MA: Harvard University Press, 1967).

Rolt, L.T.C., *Tools for the Job: A History of Machine Tools to 1950* (London: HMSO Publications Centre, 1986).

Rosenberg, Nathan, "Technological Change in the Machine Tool Industry," *The Journal of Economic History* 23:4 (1963), pp. 414–443.

Rosenberg, Nathan, *Inside the Black Box: Technology and Economics* (Cambridge: Cambridge University Press, 1982).

Rubinger, Richard, "Education: From One room to One System," in Marius B. Jansen and Gilbert Rozman (eds.), *Japan in Transition: From Tokugawa to Meiji* (Princeton: Princeton University Press, 1988), pp. 195–230.

Rule, Paul, "Towards a History of the Chinese Rites Controversy," in David E. Mungello (ed.), *The Chinese Rites Controversy: Its History and Meaning* (Nettetal: Steyler Verlag, 1994), pp. 249–266.

Shishido, Toshio, "Japanese Industrial Development and Policies for Science and Technology," *Science*, New Series 219:4582 (1983), pp. 259–264.

Smith, Crosbie, *The Science of Energy: A Cultural History of Energy Physics in Victorian Britain* (Chicago: Chicago University Press, 1998).

Smith, Thomas C., *Political Change and Industrial Development in Japan: Government Enterprise 1868–1880* (Stanford: Stanford University Press, 1968, reprint of the 1955 edition).

Spector, Stanley, *Li Hung-chang and the Huai Army: A Study in Nineteenth-century Chinese Regionalism* (Seattle: University of Washington Press, 1964).

Sugimoto, Masayoshi and Swain, David L., *Science and Culture in Traditional Japan* (Rutland, Vermont: Charles E. Tuttle Company, 1989).

Sun Yutang 孫毓棠, *Zhongguo jindai gongyehsi ziliao diyi ji: 1840–1895* 中國近代工業史資料第一輯, 2 volumes (Beijing: Kexue chubanshe, 1957).

Teng, Ssu-yü, and Fairbank, John King, *China's Response to the West: A Documentary Survey, 1839–1923* (Cambridge, MA: Harvard University Press, 1954); and John Rawlinson, *China's Struggle for Naval Development* (Cambridge, MA: Harvard University Press, 1967).

Toyoda, Toshio, (ed.) *Vocational Education in the Industrialization of Japan* (Tokyo, United Nations University Press, 1987).

Tylecote, R.F., *A History of Metallurgy* (London: Institute of Materials, 1992).

Waley-Cohen, Joanna, "China and Western Technology in the Late Eighteenth Century," *American Historical Review* 98:5 (1993), pp. 1525–1544.

Wang Jingyu 汪敬虞, "Shijiu shiji waiguo qinhua qiye zhong de huashang fugu huodong 十九世纪外国侵华企业中的华商附股活动," in Huang Yiping 黄逸平, *Zongguo jindai jingjishi lunwen xuan* 中国近代经济史论文选 (Shanghai: Shanghai renmin chubanshe, 1985), pp. 193–257.

Wang Xinzhong 王信忠, "Fuzhou chuanchang zhi yange 福州船廠之沿革," *The Tsing Hua Journal* 清華學報 8:1 (1932), pp. 1–57.

Wang, Yangzong, "A New Inquiry into the Translation of Chemical Terms by John Fryer and Xu Shou," in Michael Lackner, Iwo Amelung, and Joachim Kurtz (eds.), *New Terms for New Idea: Western Knowledge and Lexical Change in Late Imperial China* (Leiden: Brill, 2001), pp. 271–283.

Woodbury, Robert S., *Studies in the History of Machine Tools* (Cambridge, MA: The MIT Press, 1973).

Wright, M. T., "On the Lift Pump," *History of Technology* 18 (1996), pp. 13–37.

Xiong Yuezhi 熊月之, *Xixue dongjian yu wan Qing shehui* 西学东渐与晚清社会, revised edition (Beijing: Zhongguo renmin daxue chubanshe, 2011).

Yamamura, Kozo, "Japan 1868–1930: A Revised View," in Steen Tolliday, *The Economic Development of Modern Japan, 1868–1945: From the Meiji Restoration to the Second World War* (Cheltenham: Elgar, 2001) vol. 2, pp. 168–187.

Yamamura, Kozo, "Success Illgotten? The Role of Militarism in Japan's Technological Progress," *The Journal of Economic History* 31:1 (1977), pp. 113–135.

Zhang Guohui 张国辉, *Yangwu yundong yu Zhongguo jindai qiye* 洋务运动与近代企业 (Beijing: Zhongguo shehui kexue chubanshe, 1979).

Zhang Hao 張灝, "Zai chuantong yu chuangxin zhijian: shijiu shiji de zhongwen huaxue yuansu minci 在傳統與創新之間:十九世紀的中文化學元素名詞," *Huaxue jiaoyu* 59:1 (2001), pp. 51–59.

Zhang Zigao 张子高, Yang Gen 杨根, "Cong *Huaxue chujie* he *Huaxue jianyuan* kan Zhongguo zaoqi fanyi de huaxue shuji he huaxue minci 从化学初阶和化学鉴原看中国早期翻译的化学书籍和化学名词," in Yang Gen 杨根 (ed.), *Xu Shou he Zhongguo jindai huaxue shi* 徐寿和中国近代化学史 (Beijing: Kexue jishu wenxian chubanshe, 1986), pp. 105–118.

Zhongguo shixuehui 中國史學會 (ed.), *Yangwu yundong* 洋務運動, 8 volumes (Shanghai: Shanghai renmin chubanshe, 1961).

CHAPTER 2

Discovering Steam Power in China, 1828–1865

Before they came into contact with any steam engine, it can be argued, that the Chinese people had been exposed to the elementary principles which were employed in the building of that machine. They were familiar with the zoetrope, in which rising hot air generated by a burning candle turned vanes at the top of the paper lantern. In the 1670s, the Jesuits at the court of the Kangxi emperor (r. 1661–1722) also produced a model steam carriage and a model steam paddle-boat for his pleasure. The two models did not involve the use of a piston, for the carriage gears or paddle-wheels were turned by a jet of steam. Indeed, Chinese craftsmen also had a version of the piston, employed in the bellows used in furnaces. Those pistons were made of wood and driven by hand. None of that, however, added up to a steam engine, with the precision needed in fitting the piston to the cylinder, the gears that transformed a linear thrust to a rotary motion, and the sheer magnitude of the driving force that had to be served not on a lubricated metal and not a wooden frame. Until he or she came into contact with a steam engine, no Chinese person had imagined that any such machine was possible. The learning exercise had to begin with realizing that steam could be a major motive force.

In hindsight, the historian could argue that in the late eighteenth century, the Chinese elite might have had the chance to learn about the steam engine before they were presented with it, in shock, during the Opium War. In 1793, Lord Macartney, Britain's emissary to China, took with him scientific instruments and technological contrivances, including a model of James Watt's steam engine, as gifts to the Qing imperial court. Yet, he failed to present the model engine to the emperor. Nor did his mechanic-mathematician, James Dinwiddie show it to his Chinese audience in Guangzhou when he gave a lecture there about science. Had the Chinese been told of the workings of the steam engine at this early stage, history might have taken a different turn. They did not, and, thus, it is to history, rather than the counter-factual, we must now turn.

SMOKE AND WHEELS: MISUNDERSTANDING THE STEAM ENGINE

The earliest Chinese account of the steamship was written by Wang Dahai, a Chinese man who resided in Batavia, then the capital of the Dutch Indies, from 1783. In his *Haidao yizhi* (*Leisured Gazetteer of Islands*, printed in 1791), he wrote about his knowledge about the multi-racial Dutch colony and its flourishing international trade,. He included a short description of the steamship, describing steamship's machinery was driven by fire. We do not know if Wang had really seen any steamers, because the first steamer for river navigation was invented in the United States only in 1778. It is unlikely that by 1791 Wang could have seen a seafaring steamer in Batavia. It is more likely that he learnt about the technology from hearsay.[1]

The Chinese people only had the chance to witness the steamship in action for the first time when the British steamships reached their coast. They saw smoke billowing from their funnels. Whether or not they had the opportunity to read Wang Dahai description, they concluded that it was fire, not steam, that drove the engine. When, in March 1828, the first steamship from Bengal visited the city of Guangzhou in south China, the only port where foreigners were allowed to trade and reside at. An anonymous Chinese observer left the following account:

> Early in the third month [between mid and late March 1828], there suddenly came from Bengal a *huo lunchuan* [fire-wheel ship]. ... The *huo lunchuan* has an empty copper cylinder inside to burn coal, with a machine on the top. When the flame is up, the machine moves automatically. The wheels on both sides of the ship move automatically too. It can travel at a speed of one thousand *li* [about 621 miles] in a day and a night. It took only thirty-seven days from Bengal to Guangzhou. According to the foreigners the steamship was invented early in the Daoguang reign [1820–1850], but it cannot be used for shipping cargo; it is only good for carrying urgent messages.[2]

The man might have boarded the ship and saw the running of the steam engine. Yet, he missed the steam mechanism; he saw only the flame coming up and the machine moving automatically.[3] The first correct textual description of the steam mechanism appeared in the *Maoyi tongzhi* (*Treatise on Commerce*), a book on commerce in Western countries by the Prussian missionary Karl Gützlaff, who was active in the southeastern China coast since the 1830s. In that book, Gützlaff told his readers that Westerners, after observing the extraordinary force created by thermal expansion, thought of exploiting such a force by building a long iron cylinder with a rod in the middle, allowing steam entering the cylinder to drive wheels. On the same principle, Westerner built the *huo lunzhou* (fire-wheel boat) with a boiler in the middle, which generated steam and drove the wheel through a cylinder. Hence the boat could sail thirty *li* (about 18.6 miles) in an hour, travelling between the five continents in weeks.[4] The book was first published in the British colony of Singapore. We are not sure how many copies were circulating beyond colony. Nor do we know how much impact Gützlaff's description would have made on his Chinese readers.

In June 1840, the Imperial Commissioner Lin Zexu, while enforcing the opium ban in Guangzhou, reported to the throne that three *che lunchuan* [wheel ships] had reached the port of Guangzhou, he wrote that it was the heat of the flames that drove the vessels.[5] In August, when the British fleet attacked Zhejiang province, up the east China coast, an anonymous person recorded that:

> There were two wheel ships with a funnel. The ship has two masts. Its length is more than twenty odd *zhang* [more than 218 feet] and width more than two *zhang* [more than 21.8 feet]. … The funnel ship has two wheels hung outside its broadsides. The ship's cabin stores a square furnace under the beam from which the wheels are hung. When the fire is burning in the furnace, the two wheels turn like a fast mill and the ship cruises as fast as if it is flying, regardless of the wind's direction.[6]

Both observers connected the furnace to the wheels, but missed the steam mechanism which came in between. As discussed above, they would have easily referred the wheeled vessels to the historical term *lunchuan*, and, understandably, they called the steamship *huo lunchuan* (fire-wheel ship). The term "fire-wheel boat" could have derived from the ancient technology the *lunchuan* (wheel boat) or *shui lunchuan* (water-wheel boat), which used a treadmill mechanism to drive paddle-wheels. As Joseph Needham has shown, the nineteenth-century Chinese literati would have related the invention of the treadmill paddle-wheel boat to the *qianli chuan* (thousand-league boat) of the fifth-century. Chinese literati who were interested in the technology would also have known from historical records that in the Southern Song dynasty (1127–1279) the military had employed treadmill wheel-boats. Such technology might still have been in use in China in the nineteenth century.[7]

This misunderstanding led Qing officials to the effort of employing treadmill paddle-wheel vessels to combat the British naval force. At the early stages of the Opium War, Lin Zexu was interested in a type of treadmill paddle-wheel boat, the *Annan yachuan* (Annan rolling boat), because he had learned that this kind of vessels were used in Annam, now Vietnam.[8] He built four boats of this type, but no one was able to operate them.[9] One of them might have been captured by the British forces.[10]

Other Qing officials also experimented with treadmill technology. In the summer of 1840, when the British attacked, Zhejiang province, Gong Zhenlin, a county magistrate of the province, built a kind of wheel boat after he saw the British steam warships. He wrote:

> In the summer of the *gengzi* year [1840], when the British invaded and occupied Zhoushan ... there [on the coast] I saw the enemy sails standing like a forest, and among them were ships which stored fire in a cylinder and churned the water with wheels. People marvelled at their strangeness and wondered at their being powered by fire. I had an idea to imitate the wheel boat, but using hand-power instead of fire...After a few months, several wheel boats were completed. They were very convenient for going out to sea.[11]

In late 1840, when the Daoguang emperor (r. 1820–1850) dismissed Lin Zexu from the post of governor-general of Guangdong and Guangxi and sent him to Zhejiang province to assist in coastal defence, Lin brought several ship plans with him and showed them to Wang Zhongyang, a county magistrate. One of those plans showed the design of the *Annan yachuan*.[12] In 1841, Wang was laying down plans for building similar vessels. According to his description, these boats had two wheels in the fore compartment and two in the aft, each wheel having six blades. Within the ship two men stood shoulder to shoulder, pushing or treading the treadmill.[13] Those boats were probably captured by British forces when they were near completion. They thought the Chinese were imitating their steamships.[14]

There were similar efforts being made in building treadmill paddle-wheel vessels in other parts of coastal China. In Guangdong, in 1841, Chang Qing, a junior official in Guangzhou, built a treadmill paddle-wheel ship.[15] Unfortunately, we have only the drawing of it but not the technical details (Fig. 2.1).

A story about a Chinese man's attempt to build a bogus steamship shows that the Chinese people were struggling with the idea of how to drive paddle-wheels. In 1841, a Ningbo man who worked as a domestic

Fig. 2.1 Chang Qing's wheel boat. (Source: Joseph Needham, *Science and Civilisation in China*, volume 4, part 2, (Cambridge University Press, 1965), figure 638

servant claimed to roughly know the mechanism of the steamship. He made a model out of wood and bamboo with a lighted candle in the middle. This model ran on the water and attracted much attention. The governor-general of Jiangsu, Jiangxi, and Anhui recruited this man to build his version of the *huo lunchuan*. The building work dragged on, the man blaming his failure on the lack of workmen. The project was finally abandoned when the British attacked Zhejiang.[16]

In the later part of the First Opium War, officials in Jiangsu province, just north of Zhejiang, also built treadmill paddle-wheel boats to combat the British naval force. In June 1842, in a battle on the Wusong River (also known as Suzhou Creek in English literature) that passed through Shanghai county. They were amazed by the Chinese ingenuity:

> [16 June, 1842] Some way further up the [Wusong] river, fourteen war-junks were in sight, and also five large newly-built wheel-boats, each moved by four wooden paddle wheels. ... These wheel-junks were fitted with two paddle-wheels on either side, strongly constructed of wood. The shaft, which was also of wood, had a number of strong wooden cogs upon it, and was turned by means of a capstan, fitted also with cogs, and worked round by men. The machinery was all below, between decks, so that the men were under cover. They were all quite newly-built, and carried some two, some three, newly-cast brass guns, besides a number of large ginjals [heavy muskets].[17]

Another British officer recorded:

> A native of Chusan [Zhoushan] had built small vessels on the model of our steamers with paddle-wheels. It was said that, when ready, he endeavoured to propel them by means of smoke made in the hold; but, as they declined altogether to move on such terms, it was subsequently found advisable to turn the wheels by relays of men working with their weight, somewhat on the principle of the treadmill. In this condition they were found by our force. The Chinese officers tried them, and were satisfied with the rate of their movement, and the use of the guns on board.[18]

The idea of using smoke is absurd. Yet, this anonymous inventor must have been trying to work out the link between the smoke he saw from the steamships and the Chinese traditional treadmill technology. Unfortunately, all those wooden-structured vessels that were fitted with manual treadmill mechanisms were not comparable to British steamships. At the end of the

Opium War in 1842, Qing officials still had no idea of the mechanism of steamships. Qiying, the imperial commissioner who signed the Treaty of Nanking, boarded a British steamer and witnessed the machinery. He reported what he saw to the Daoguang emperor:

> The length of the fire-wheel ship your slave [I] boarded is about five *zhang* [about 54.6 feet] and the width of it is half of the length. The ship is fitted with a water and a fire cylinder. When coal is burnt [in the fire cylinder], fire flares and smoke rises. Both inside and outside [of the water cylinder] there are gears, which are agile. [The fire cylinder] is roughly based on the principle of clock. Hence [the ship] can cruise fast without the sail. It is rumoured that there are men or oxen driving the gears. But this is speculation.[19]

The description shows that Qiying misunderstood the mechanism. He found it incomprehensible that the wheels could have been turned without an animal-driven motive force.[20] Unable to reproduce wheeled steamboat, Chinese officials sought foreign expertise. A Guangzhou merchant, Pan Shirong, financed the building of a steam paddle-wheel boat with the assistance of a European technician at Guangzhou in the spring of 1842.[21] Unfortunately, we have no description of this boat. The emperor was interested and ordered officials in Guangzhou to find out the construction cost. They reported that Pan's ship did not run very well, but added that there were foreign technicians who were capable of building steamships at Macao at a high cost. They asked if they were to build or to purchase from the Westerners.[22] Obviously not convinced, the emperor ordered them not to hire foreign technicians or purchase any steamships on the grounds that these ships were not useful.[23]

However, a development after 1842 shows that Qing officials in Guangzhou did make further attempts to build the steamer. Foreign missionary records show that, by April 1845, Qiying had recruited an American shipwright in Macao to build a steamship.[24] Yet, its building work was hampered by the shipwright's death in 1847.[25] We have no more details about this steamer. Whether or not the shipwright had left any plans or written instructions, the Chinese workmen, who were most likely worked from an oral tradition, could not continue the construction work. Hence, the project had to be suspended after the shipwright died. There was no sign yet that a breakthrough had come, and when it did, it did not herald the success that historians made it out to be.

There is also the curious story of the Chinese man who was connected with foreign trade in Guangzhou and who met the British missionary, George Smith in April 1845. He asked if Smith could "furnish him [with] a diagram and [a] explanation of the manner in which foreigners could weave and manufacture cloth by steam-machinery..." Smith gave him a lecture about the application of the steam engine. There simply is no record of what happened after the event. Chinese knowledge of the steam engine was extremely sketchy at least up to this time.[26]

The Few Who Understood the Steam Mechanism

The first Chinese we know of who tried to build a steam engine was Ding Gongchen, a Chinese Muslim from Fujian province. We know very little about his life. In the 1830s, he did business in Southeast Asia. By 1841, he was residing in Guangzhou. He might have had access to foreign books about Western technology circulating in Batavia, Macao or Guangzhou at the time. He was mostly interested in artillery technology, and in 1841, he first printed the *Yanpao tushuo* (*Illustrative Treatise on Gunnery*), which was mostly about gunnery and gun casting but also included an essay discussing his experiment of building a model steam engine. According to it, Ding saw a model steam engine in Guangzhou and figured out its mechanism. He asked artisans to build a small locomotive, which was about nearly 2 feet (1 *chi* and 9 *cun*) long and 7.5 inches (6 *cun*) wide. It was capable of carrying a load of around 39 lbs. (30 catties).

His book provides two diagrams, one of the cylinder (Fig. 2.2), and one of the locomotives (Fig. 2.3), which show how his model engine might have looked. According to the historian David Gwyn, Ding's model locomotive was actually a combination of different engine designs, which could be found in Britain in the early nineteenth century. Th engine is a 0-2-2 wheel arrangement, which was first used by Robert Stephenson, a British locomotive pioneer, who developed the innovative *Rocket*. Ding's design of vertical cylinder is similar to the stationary table engine, which was developed by the British engineer and inventor Henry Maudslay in the 1800s. Ding's vertical boiler, which is positioned in the middle of the frame, is in part cylindrical and in part domed-shaped. In fact, the vertical boiler was quite different from the horizontal barrel type boiler—which had been a more common style in early steam locomotives, but was similar to a fire engine design developed in the 1820s. In the vertical-style boiler, the firebox is on the bottom and the domed-shaped top is where steam

2 DISCOVERING STEAM POWER IN CHINA, 1828–1865   31

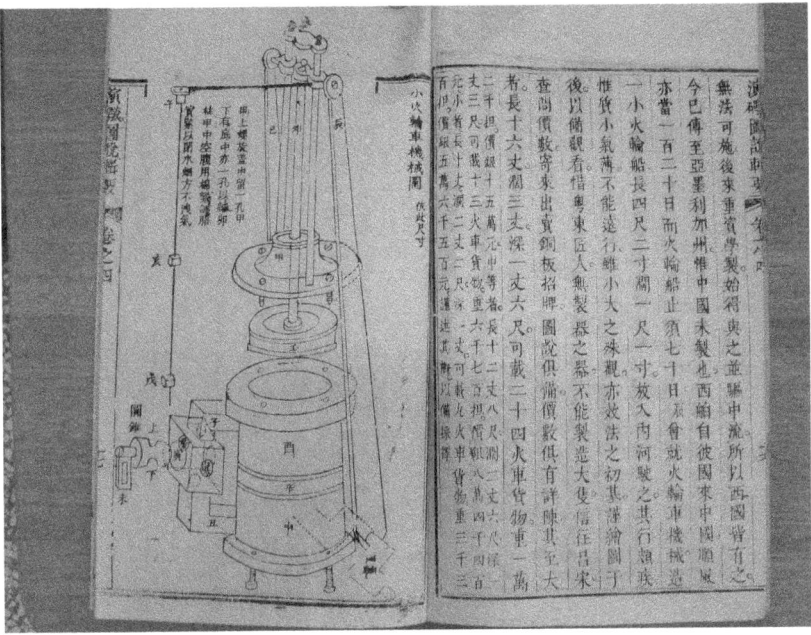

Fig. 2.2 Ding Gongchen's steam cylinder. (Source: Ding Gongchen, *Yapao tushuo jiyao* (1945 reprint of the 1843–1851 edition, kept in the Xiamen University Library))

gathers; the segmental shape of the smokebox suggests that there was a horizontal barrel that housed fire tubes leading to the smokebox. Because the steam exhausts directly from the cylinder through the plug-valve, the smokebox (which was in the front of the locomotive, under the chimney) did not allow the steam to travel through the blast-pipe. This design increased the engine's efficiency because it drew hot gases from the firebox through the boiler tube. In a further development of the 1830s, the driving cranked axle wheel was included in a locomotive design.[27]

Furthermore, Ding's *Yanpao tushuo* gives a detailed explanation about how his model engine could have worked. Ding writes that, after one fills the boiler up to the safety valve with water and then closes all of the boiler's inlets and outlets, one would then burn coal or wood in the firebox to heat up the water (Fig. 2.3). Once the water is boiling and generating steam (*qi*), one then switches the steam regulator on, which allows steam

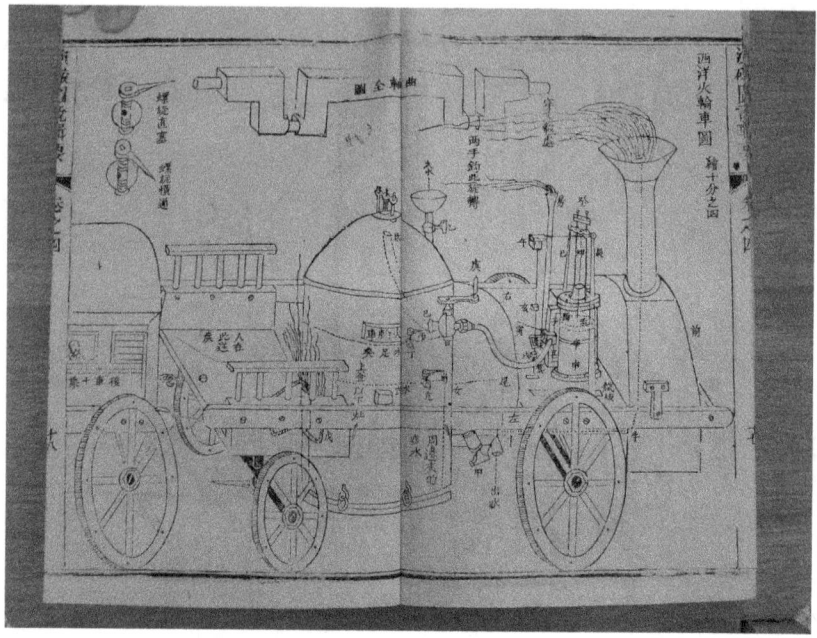

Fig. 2.3 Ding Gonchen's steam locomotive. (Source: Ding Gongchen, *Yapao tushuo jiyao* (1945 reprint of the 1843–1851 edition, stored in the Xiamen University Lirbary))

to enter the cylinder through a pipe. Additionally, the cylinder (Fig. 2.2) has a piston in the middle (the piston is packed with cotton-cloth and is well-greased). On the side of the cylinder, a conically shaped plug-valve that regulates the steam's direction is housed in a square casing, and the valve's lever has a rectangle hole, through which the valve is connected by an iron-wire with the cross-head at the top of the piston-rod. The cross-head itself is attached on each end to two side rods, which—through passing down on either side of the cylinder to below the iron frame of the locomotive—are connected to the cranked axle wheel. As for the plug-valve, it has two grooves, one on the top and the other the bottom; the boiler's steam pipe is connected to the inlet on the right-hand side of the drain cock.

Another interesting feature described in Ding's *Yanpao tushuo* is that steam (*qi*) from the boiler enters the lower part of the cylinder through the lower groove of the valve, forcing the piston and the valve lever up simultaneously. When the piston reaches the top of the cylinder, the plug-valve would be pulled up by the iron-wire upward by roughly 45 degrees, and the steam would enter the upper part of the cylinder through the valve, forcing the piston down. Then the iron-wire would pull the plug-valve downward, about 45 degrees. Hence, the piston moves reciprocally, while the two rods that connect the beam and the piston-rod drives the cranked axle. Finally, the steam would have completed its work on the piston through the plug-valve and then the pipe. Ding's book also explains that the engine does not have a brake mechanism; thus, human force would be required to turn the cranked axle forward or backward. Clearly, Ding understood that it was steam that drove his engine, even though his writing does not discuss ideas similar to steam pressure.

Ding Goncheng's experiment was a success. He used the same engine to build a model paddle-wheel steamboat (which was about 4.4 feet [4 *chi* and 2 *cun*] in length and 1.15 feet [1 *chi* and 1 *cun*] in width), which he then put to the test on the Pearl River. (Unfortunately, Ding's book does not describe the model steamboat.) Overall, Ding was excited and knew that his experiment marked the beginning of his having acquired important Western skills and knowledge. However, because Ding lacked proper tools, building an efficient engine was difficult for him. While describing his model steamboat, Ding wrote:

> It runs with good speed, but on account of the fact that the boat is small and the steam is weak it cannot go far. Though the model is small it marks the beginning of our effort to imitate the Western method. ... Unfortunately the craftsmen in Guangzhou possessing no tools that build machines, cannot build big ships.[28]

Although we are not sure whether Ding meant machine tools, he certainly knew he needed proper tools to build a full-scale steamer. Notably, Ding Gongcheng's text and illustrations are rich in details. Both Figs. 2.2 and 2.3 follow traditional drawing conventions, but they also show some Western conventions, such as scales and dotted lines. It is also important to note that Fig. 2.3 fails to provide enough details for the plug-valve, which is crucial for the reader to understand how the steam's direction might be regulated. As a result, readers are left to wonder about the firebox's exact position and shape, and about how the smoke might have come out of the funnel (although the text provides a rough idea, and the diagram indicates it with a dotted line from the discrete furnace to the top of the funnel). Readers would have had to speculate about the exact mechanical arrangement under the frame, although a separate drawing of a crankshaft gives a rough idea about how it might work.

Nevertheless, the difficulties in pictorial representation did not stop Ding's knowledge from circulating in the aftermath of the First Opium War. In August 1842, the Daoguang Emperor demanded that local officials present him with a copy of the *Yanpao tushuo*. The Emperor also demanded a report on Pan Shirong's steamer building project.[29] Zheng Fuguang, a literatus who was from Anhui province in eastern China and was interested in natural philosophy, mathematics, and optics, took effort to study the steam mechanism.[30] By 1846, Zheng had written a book about optics, in which he included two essays on the steam engine and steamboat, "*Huo lunchuan tushuo* (Illustrative Treatise on the Fire-wheel boat)," and "*Dingbu huolunchuan jiju tushuo zhiwu* (Correcting the errors in the Illustrative Treatise on the Fire-wheel boat)," respectively.[31] According to the former, Zheng had seen an illustrated explanation of a steamboat, but he could not understand the mechanism. Later, however, he did encounter a model steamboat, which was about 5 or 6 *chi* (6.6 feet) long and had an engine that was exposed on the deck. Afterward, while being in China's capital city of Beijing, he received diagrams of the steam engine from Ding Shoucun, who was a low-ranking government official that was interested in military technology, especially explosives. Zheng had read Ding Gongcheng's *Yanpao tushuo*, and he had also discussed the mechanism of Ding's plug-valve and cylinder, although he had misunderstood the mechanism. In an attempt to gain a better understanding, Zheng communicated with Ding between February and March 1847; in the course of their correspondence, Ding sent his model plug-valve to Zheng. Thereafter, Zheng wrote the second essay, which corrected his errors in the previous essay.[32] However, Zheng did not conduct any experiments based on his study; rather, his diagram of the steam engine is actually a crude copy of Ding's illustrations and offers even fewer technical details—although Zheng does explain the steam mechanism in his text (Fig. 2.4).

In addition to Ding's and Zheng's work, there were other publications about the steamship in circulation—especially in Guangzhou. Liang Tingnan, a Guangdong literatus who had participated in coastal defense during the First Opium War, was interested in Western cultures and histories and, by 1846, wrote a booklet about England titled *Lanlun oushuo*

Fig. 2.4 Zheng Fuguang's steamboat diagram. (Source: Zheng Fuguang, *Jingjing lingchi* 3:37b (Taipei: Yiwen yinshuguan, 1966, reprint of the nineteenth century edition))

(*Occasional Discussion on England*), which contains a short section about the steamship. *Lanlun oushuo* describes that the expansive property of steam could be used to drive locomotives and steamships (*huo lunchuan*), even though it does not explain how steam might be turned into motive force. We do not know, however, that Liang did, in fact, acquire such knowledge.[33]

Because knowledge of the steam engine was transmitted through texts and illustrations, if any Chinese literati or artisans had wanted to reproduce Ding's model steam engine, they would have had to go through all the same trials and errors that Ding Gongchen had undergone to ensure that the shapes and mechanisms of all the components worked correctly. Thus, lacking any better means to accurately and repetitively convey technical details of technology, it was difficult for the new technology to diffuse throughout China, let alone develop.

However, despite the difficulties in knowledge transmission, efforts were made to reproduce the novel technology. One such example comes to us from Aurelius Harland, a British medical missionary. According to Harland's account, in 1847, he visited the prominent Guangzhou merchant Pan Shicheng, who during the First Opium War had already contributed money to building four ships according to the Western model. Harlem recorded that, in Pan's garden, he saw a steamboat built by an anonymous Chinese man who commanded good English.[34] Harland recorded that:

> The boat itself is 70 feet in length and 15 feet beam, built exactly in the Chinese form as no other would yet be allowed—all the machinery including boilers etc. is on deck, the paddles are principally of wood 10 feet in diameter and about 2 ½ in width, very beautifully made, and strongly fastened with iron; there is only one cylinder of a foot diameter inside and a 3 feet stroke. It is laid on deck horizontally, but being nearly two feet below the level of the paddle shaft will work with great loss of power, for when the cam is at the highest point there will be a large portion of portion [sic, power?] wasted in the tendency to tear up the beams which contain the groove in which the cross bar at the head of the piston slides. The boilers are 4 in number, placed side by side in one furnace, each of them is 12 feet long and 2 ½ feet in diameter, with a fire flue of 8 ins through each. They are made of brass hooped with iron, and the brass plates are brazed instead of being riveted together. As there is only one cylinder and no fly wheel, the motion will be very intermitting [sic], and the starting will be very difficult for there is nothing but the momentum of the vessel to carry the crank past the line of centre when the piston reaches each extremity of the stroke. The man is very intelligent and speaks good English—I pointed out some faults and improvement[s] which he said he would adopt saying he hardly expected to succeed at first but hoped to make a good one after 3 or 4 attempts when the workmen had a little more experience—he expected to have all this completed about a fortnight from this time.[35]

At 70-feet in length, and having a 15-foot beam, this was the full-scale steamboat that Ding Gongcheng could not achieve. (Notably, it remains unclear how the cylinder-piston and the boilers were built.) The gearing of two 10-feet-in-diameter paddle wheels, a one-foot-long cylinder, and a set of four boilers would have required engineering competence. Unfortunately, Harland's description does not tell us what faults in the arrangement he had seen. Nor do we know whether the engine worked after Harland offered his advice for improvements.The efforts of Ding Gongcheng, alongside the anonymous Chinese shipwright in Pan Shicheng's north Guangzhou garden, show that steamboat technology remained difficult to reproduce in China in the 1840s. Nevertheless, pockets of interested people were talking about it, and they were attempting to come to grips with it in terms of their own knowledge-base.

The writings, such as those of Gützlaff and Zheng Fuguang, were included in Wei Yuan's *Haiguo tuzhi* (*Illustrative Gazetteer of Maritime Nations*), a collection of translated essays on world geography. These texts provide some important information about foreign and domestic technology that could be used for maritime defense. It is against this background that of trial and error in using indigenous expertise and knowledge to emulate the steamer that the early appeal for hiring foreign technicians and importing Western tools. As such, we see that important Qing government officials launched more experiments in the early 1860s, and that they did so within a changing technological environment.

China's Changing Technological Environment

The First Opium War was concluded in 1842 with the signing of the Treaty of Nanking. As part of the peace settlement, the Qing government agreed to open five ports in which foreign merchants were permitted to trade and reside. This allowed both skills and knowledge about steamship technology to enter China via increased trade and more contact between foreigners and the Chinese, despite the fact that the Qing government did not make an immediate effort to assimilate steamship technology.

After the 1840s, low-level engineering skills were introduced into the treaty ports because of the increasing number of merchant steamships cruising along China's coast. For example, a British shipping firm, Peninsular & Oriental Steam Navigation Co., opened its regular line between London and Hong Kong, the new British colony, in 1845.[36] This, in turn, caused dockyards for servicing and building steamships to emerge in the treaty ports. In the late 1840s, a foreign dockyard was established in Shanghai, and the number quickly rose to six in the 1850s and then to nine in the 1860s.[37] By 1860, three docks were operating on Hong Kong Island and four in the port of Huangpu, near Guangzhou. After 1860, with more treaty ports having been opened throughout China, more dockyards were built to meet the demands of the growing shipping business.

Those dockyards were well equipped and even had the capability to build steamers, including their engines. For example, the British dockyard firm Boyd & Co. in Shanghai had installed steam engines and machine tools such as lathes, planers, boring mills, rolling mills, a steam hammer, and a furnace. In 1865, the dockyard was hailed as one of the best equipped in the Far East.[38] Three years after the firm was established in 1862, Boyd & Co. built 17 small, 70-horsepower steamboats. In 1870 the firm built a 1300-ton steamship, including its steam engine and boiler.[39] According to *The North China Herald*, a weekly English newspaper published in Shanghai, in 1880, Boyd & Co. employed 1000 to 1400 Chinese workers, all of whom had the ability of executing the blueprints to which they were assigned.[40] Additionally, an American dockyard, Farnham & Co. (established in 1864), hired more than 2000 Chinese workers in the 1880s.[41] In other treaty ports, such as Xiamen, Fuzhou, Tianjin, and Yantai (known to foreigners as Chefoo), foreign-owned dockyards and machine shops were also very active.[42] Although we lack descriptions of these Chinese workmen, the machine shops would have trained them with modern engineering skills, including the ability to read technical drawings and operate machine tools.

Furthermore, after the 1840s, Christian missionaries in Hong Kong and the treaty ports started to publish newspapers and journals, which introduced Western learning as a means of proselytizing. For example, from 1849 onward, a missionary medical doctor Benjamin Hobson published a series of books that introduced knowledge on human anatomy, astronomy, and natural science. Among them, the volume on natural science, *Bowu xinbian* (*Natural Philosophy*, or literally *The Broad Learning of Things Newly Compiled*), which was first published in 1855 in Guangzhou and reprinted in Shanghai by the London Missionary Society Press, contains sections about the atmosphere and steam engine mechanism.[43]

The opening essay of the *Bowu xinbian* is "On the Atmosphere" (*Diqi lun*), and it introduces basic scientific ideas about the atmosphere.[44] The Chinese term *diqi*, which literally means the air of the earth, is a fitting translation for the English word atmosphere. Hobson's essay states that the earth is a round object surrounded by air (*qi*), and humans need air like fish need water.[45]

Although Hobson's text does not invent a new term by which to translate atmospheric pressure, it does offer that the weight of the atmosphere is 15 pounds per square inch, which indirectly suggests the idea of pressure. In "On the Atmosphere," Hobson also writes that one can use an instrument called a *fengyu zhen* (i.e., a barometer; the Chinese literally means "wind and rain pointer") to measure it. Nor does Hobson invent a new term to denote the word *vacuum*; but he does give a description of how to use an air pump (i.e., *qiji tong*, which literally means "steam engine cylinder") to pump air out of a glass vessel, in which small animals or plants cannot survive, nor could a fire continue to burn. Additionally, the text describes several experiments about the vacuum to explain how the atmospheric pressure exerts a force on the vessel in which a vacuum is created by the air pump.[46]

Furthermore, in the chapter titled "On Steam" (*Zhengqi lun*), Hobson explains the expansive property of steam (*fasan zhili*) and how steam pressure (i.e., *qili*; literally translated as "the force of steam") could be turned into a motive force. The chapter explains that, when heated and vaporized, one square inch of water expands 1700 times, which then produces pressure within the cylinder. The higher the temperature, the higher the pressure. Of course, the steamboat and the steam locomotive both, utilize such a force with its engine on board, thus empowering it to travel both quickly and far. Finally, in this chapter, Hobson also provides a detailed description, including illustrations, of the Watt engine.

In the same period of time, a missionary organization, Morrison Education Association, in Hong Kong, published a magazine titled *Xia'er guanzhen* (*Chinese Serial*), which contained articles on Western politics, earth science, and technology. In September 1853, the magazine published a short essay on the construction of the steam mechanism entitled "*Huochuan jizhi shulue*" ("A Rough Description of the Mechanism of the Fire-Wheel Boat"), which explains the mechanism of the steamboat in terms of steam and the atmospheric pressure. The essay also introduces the air pump, which can create a vacuum in a glass canister—just as Hobson's writing had done. Specifically, "*Huochuan jizhi shulue*" explains that both steam and air have elastic properties.[47]

In contrast to Ding Gongchen's and Zheng Fuguang's writing on the steam engine, *Xia'er guanzhen* and *Bowu xinbian* did, in fact, give their

Chinese readers basic knowledge about how steam pressure might have driven the steam engine. Below, we will see how writings might have had an impact on China's first effort to build the steamer.

The First Chinese-built Steamer

While new knowledge and engineering skills started to enter China, Qing officials' use of Western technology was increasing during the Taiping Rebellion. Adding to the demonstration of Western fire power was another foreign expedition against the Qing dynasty in the Second Opium War. After taking Guangzhou in 1857, the British-French allied fleet steamed north. They seized the city of Tianjin and then the capital Beijing in 1860. The imperial court was forced to flee. The British-French allied force retreated after the signing of the Peking [Beijing] Convention, but the imperial court had, by then, witnessed Western fire power at first hand.

The Chinese officialdom was increasingly interested in Western arms and steamers. However, before any steamers could be built under official direction, the same trial and error experienced by Qiying and Ding Gongchen had to be repeated. Fortunately, with a changing technological environment, the new experiments led to a more fruitful consequence.

In December 1861, after the strategic city Anqing in Anhui province was retaken from the Taipings by Qing troops, Zeng Guofan, the leading Qing military commander and newly appointed governor-general of Jiangsu, Jiangxi, and Anhui, established an arsenal for producing ammunitions for his troops in that city.[48] Zeng wanted not only to manufacture ammunition for Western-type firearms, but also build steamers. In 1861 Zeng Guofan recruited Xu Shou, Hua Hengfang, and others who had some knowledge of Western science and technology to join his staff. Xu was interested in mechanical engineering and Hua in mathematics. Yet, no document shows that Xu and Hua had learned Western drawing conventions. Nor do we know how much engineering capability they had. Under Zeng's instruction, they began to experiment with steamship technology. Yet, they still faced the same difficulties Ding Gongcheng had twenty years earlier.

In July 1862, Xu Shou and Hua Hengfang built a model engine. The diameter of the cylinder was only 1.70 inches.[49] It was presented to Zeng Guofan, who described the running of the engine in his diary entry of 30th July:

> Hua Hengfang and Xu Shou brought the engine of the fire-wheel boat which they had made here for a demonstration. The method is to use fire to make steam, and direct the steam into a cylinder which has three holes. When two of the front holes are closed, the steam goes into the other front hole. The piston automatically goes backward and the wheel turns the upper half circle. When two of the back holes are closed, the steam goes to the other back hole; then the piston automatically moves forward and the wheel completes the rest of the circle. The bigger the fire, the greater the quantity of steam. The engine moves forward and backward as if it were flying. This demonstration lasted for an hour. I was so happy that we Chinese could make these ingenious things like the foreigners. No longer will they be able to take advantage of our ignorance."[50]

Zeng's description of the movement of the valves is incomprehensible. It is likely that Zeng did not understand the mechanism but was only describing what he had seen.[51] Yet, he had learned from Xu and Hua that it was steam that drove the engine piston.We do not have any description of how Xu and Hua built the engine, but we know how they might have acquired their knowledge. They consulted Dr. Hobson's *Bowu xinbian*. According to the 1880 reminiscences of John Fryer, a British missionary who from 1867 became Xu and Hua's colleague in a Qing government Translation Department of the Jiangnan Arsenal:

> The Viceroy [Zeng Guofan] required him [Xu Shou] to build a steam-boat, and reluctantly he consented to make the attempt. He first made a model of an engine from the somewhat rough illustrations in Dr. Hobson' work before referred to [Natural Philosophy]. This proving to be a success, he was encouraged to proceed with the more difficult task assigned him. By means of Chinese tools and materials, and such ideas as he contrived to get through looking carefully over a small steamer at Anching [Anqing], he managed to prepare his designs, and commenced his work with no foreign assistance whatever. He met with a most determined opposition from local officials, but, assisted by his son and encouraged by the Viceroy, who took a lively interest in the proceedings, the work was at length completed; not however without at least one entire failure.[52]

Apart from reading translated Western writings, Xu and Hua boarded foreigners' steamships moored in the Yangzi River to observe their engines. They also boarded the two American gunboats rented by Zeng Guofan in Anqing.[53] Presumably, they had the chance to examine the ship engines carefully, and learnt enough to build the model engine. From their model, Xu and Hua proceeded to the steamboat. After spending a year in preparing Chinese tools and drawing up plans, Xu and Hua finished a steamboat in around November or December 1863. It was a wooden-hulled boat with a screw-propeller system. It cruised for about 0.36 miles [one *li*] and ran out of steam, because they did not make boiler tubes and, hence, built a faulty engine.[54] Later in December 1868 Zeng Guofan described it in his memorial as follows:

> We used only Chinese workers and hired no foreigners at all, but though we built a steamboat and made it go, it was very slow, and we could not grasp the method [of building a steamboat]."[55]

Hence, we can tell that Xu and Hua had not quite mastered the technology.[56] In 1864, Xu Shou and Hua Hengfang moved to Nanjing after the city fell to Zeng Guofan's forces. They continued their experiment there. In April 1865, they completed a paddle-wheel steamboat and put it to test on the Yangzi River. It was able to travel about 7.16 miles (20 *li*) per hour.[57] Zeng Guofan's son named the boat *Huanghu*. According to the *North China Herald*'s report in 1868, the *Huanghu* was 25 tons and its length was about 60 feet (55 *chi*). It was equipped with a single-cylinder high-pressure engine. The cylinder was about 1.09 feet (1 *chi*) in diameter, 2.1 feet (2 *chi*) long, and the shaft was 15.3 feet (14 *chi*) long and 2.2 feet (2.04 *chi*) in diameter. The boiler was 12.3 feet (11 *chi*) long and 2.8 feet (2.6 *chi*) in diameter. It had 49 tubes which were 8.7 feet (8 *chi*) long and 2.6 inches (2 *cun*) in diameter. Most of the material used was obtained locally, only the iron used in making the shaft, the funnel, and parts of the boiler were imported. According to the *North China Herald*, all the tools and apparatus, together with the bolts, screws, valves, and pressure gauge, were made under the direct supervision of Xu Shou and his son without foreign models or assistance.[58]

Students of Chinese history have long hailed the *Huanghu* as a technological achievement of Xu and Hua's ingenuity. They might have implied that by building a workable steamer before Chinese modern shipyards were established, Xu and Hua had been able to harness steam technology using indigenous skills. Yet, judging from the *North China Herald*'s description, the *Huanghu*'s engine and tubular boiler could not have been built without machine tools.[59] Furthermore, Xu and Hua would have needed workers who knew how to produce parts and assemble them.

The point here is neither to downplay Xu Shou and Hua Hengfang's achievement nor to deny the Chinese people's capability in learning Western technology. They certainly had the intellectual capacity to understand the steam mechanism. Yet, the importance of machine tools and technical drawings in modern engineering cannot be ignored. Parts such as the cylinder, piston, copper tubes, shaft, and even small parts like valves and screws, had to be made with machine tools. Furthermore, the complexity of building a steamboat had gone beyond the writings that Xu and Hua had read. Gearing the engine, boiler, and the paddle-wheel system could not have been achieved without following some sort of design. Yet, no evidence shows how much engineering ability Xu and Hua had obtained by self-learning and experimenting.[60] Where might the tools and the workmen who could operate them to build the *Huanghu* have come from?

MACHINE TOOLS

In the campaign against the Taipings, a sustained supply of ammunition was crucial to Qing troops. This was why Zeng Guofan established his Anqing arsenal. Its workers probably used very basic traditional hand tools to produce ammunition. Both the Ever-Victorious Army, a foreign-Chinese mixed mercenary force, and Li Hongzhang's Huai Army employed Western firearms much more extensively than Zeng Guofan's Xiang Army, and to ensure his supply of ammunition, in 1863, Li set up three arsenals in Shanghai and its surroundings. One of these was directed by Halliday Macartney, a former British army physician serving under Li Hongzhang.

All three arsenals employed Chinese workers and adopted Western techniques, and Macartney's arsenal, for certain, was equipped with Western machine tools. The machine tools, nevertheless, were not in place in 1863, for Macartney's biographer noted that he produced "shells, fuses, and friction tubing" for Li Hongzhang's army "without an engine, cupola, or indeed a single tool beyond a hammer and a file" and "a melting apparatus was [extemporized] out of the clay from an adjoining field."[61] The machine tools materialised in January 1864, when Macartney persuaded Li to buy the set left over from the disbanded Lay-Osborn Flotilla. Those tools included a steam engine, a boiler, and unknown number of lathes, cupolas, and fans.[62] With this set of machine tools and more purchased from Hong Kong in 1863, Macartney and his workers, by then moved to

Suzhou, produced percussion caps, shells, fuses, and detonators in greater numbers. They even made mortars.[63] In 1864, Macartney's arsenal was able to produce 24-pound shrapnel. According to Macartney, Li was amazed at what machine tools could do when Macartney showed him the product, though he initially had doubts about the Lay-Osborn machine tools. By May 1864, the Suzhou arsenal employed four to five foreign technicians under the direction of Macartney.[64]

The technology of this arsenal approached Western standards. With the machine tools driven by a steam engine, the gun barrels were bored from solid cast iron pillars.[65] This was a clear departure from the traditional gun making techniques, which cast cannons by pouring molten iron into an earth mould without boring the cannon barrels. Although the Suzhou Arsenal could only produce mortars and shells, Li Hongzhang had learned the potential of the steam engine and machine tools in making larger guns.[66]

Zeng Guofan's ambition to build steam vessels with Chinese hands also led to his obtaining a set of machines tools. When Xu Shou and Hua Hengfang were experimenting on their first steamboat, Rong Hong (better known as Yung Wing), an American-educated Cantonese, was introduced to Zeng Guofan. He met Zeng and his staff in September 1863 to discuss the possibility of establishing a machine shop. When consulted, Rong Hong suggested a machine shop. He said:

> I would say that a machine shop in the present state of China should be of a general and fundamental character and not one for specific purposes...a machine shop that would be able to create or reproduce other machine shops of the same character as itself; each and all of these should be able to turn out specific machinery for the manufacture of specific things. A machine shop consisting of lathes of different kinds and sizes, planers, and drills would be able to turn out machinery for making guns, engines, agricultural implements, clocks, etc. In a large country like China...they would need many primary or fundamental machine shops but after they had one...they could make it the mother shop for reproducing others....[67]

In other words, Rong Hong understood well that China needed machine tools before guns, steam engines, and many other modern technological artefacts could be produced. Rong Hong's meeting with Zeng Guofan took place just after Xu Shou and Hua Hengfang had finished their model steam engine and were about to start to build their first steamboat. The experience could have made it possible for Rong Hong to persuade Zeng

Guofan that machine tools were essential in machine building. Zeng Guofan was convinced and commissioned Rong Hong to purchase a set of machine tools from the United States for 68,000 taels. Rong Hong left for the United States in October 1863. The machine tools arrived at Shanghai in 1865 and were incorporated into the new Jiangnan Arsenal (*Jiangnan jiqi zhizao zongju*, Jiangnan General Bureau of Machinery Production), which was established by Li Hongzhang in June 1865 with the purchase of a machine shop from an American firm. It built its first wooden-hulled steamers with a purchased steam engine in 1868. The Qing government also granted permission to establish the Fuzhou Navy Yard (1866), which not only built its first steamer in 1869 but also set up China's first engineering school, training both workers and naval engineers. Along with foreign owned dockyards, they became the major institution that introduced Western steam technology by the end of the nineteenth century.

Conclusion

From the treadmill paddle-wheel boat to Jiangnan Arsenal's steamers was a long journey of discovery. Qing officials experimented with knowledge and skills available to them, and it took time, and trial and error, for them to realize that steamboats were driven by steam, and that machine tools were necessary to turn the principle of steam into a workable engine and the drawings had to made and read for the technology to be diffused.

The discovery of steam technology, therefore, marked more than the discovery of a new motive force. It marked the beginning of a radical change in the methods of industrial production. It meant a shift from working with wood towards working with metals for machine parts, more complex machines made up of components of different materials, shapes, and functions, and a stricter division between the machine designer and the workman. It also marked the use of technical drawings to a very much higher standard of precision than had been achieved by Chinese artisans even with their scale models. The oral tradition which served well enough between master and disciples was no longer sufficient in the communication and transmission of knowledge and skills, for the Western technical drawings taught that workmen were able to produce and assemble parts according to instructions represented by commonly agreed upon conventions. Working with steam was to bring with it an entirely new work ethic that included more than the use of machine tools and technical drawings. Much of that new ethic would be introduced in institutions such as foreign-owned dockyards in the treaty ports and government shipyards. In a steady, although not rapid pace, they were driving China's technological changes.

However, the story of discovery was only a beginning of China's entry into modern technology. The birth of thermodynamics, which epitomized the drive for efficiency in machines such as the steam engine, took place in the 1850s. By then, driven by the need to produce more efficient engines and machines, engine builders employed mathematics to study the relationship between heat, mechanical work, temperature, and energy. It became essential in training engineers. Along with sophisticated machine tools, machine building was no longer a matter of trial and error but was defined with mathematical precision. Technology progressed in a breathtaking speed and new types of engines drove larger vessels, trains, and factory production. To catch up, China needed the institutional capacity that could absorb new knowledge and skills from the West, train massive numbers of technicians, and organize them in the production of firearms, steamers, and many other machinery. Institutional setting put technical drawings and machine tools in the heart of both the training program and the production procedure. It was not an easy task. The next chapter will discuss how China might have tackled the scientific differences between China and the West by translating the concept of heat.

Notes

1. Wang Dahai, *Haidao yizhi*, 1843 reprint of 1791 edition, kept in the Fu Ssu-nien Library, Academia Sinica, Taipei.
2. One *li* is about 576 meters. But "one thousand *li*" is an often-used metaphor to describe high speed and long distance. See Wei Yuan (comp.), *Haiguo tuzhi* (Shanghai: Shanghai guji chubanshe, 1997, reprint of 1876 edition), 52:12b. Wei Yuan (1794–1857) was an open-minded, educated elite concerned with various crises that the Qing empire was facing, especially after the dynasty was defeated in the First Opium War. He compiled and printed the first edition of the *Haiguo tuzhi* in 1843, which included translated works about the Western world and essays about maritime defence. He later extended the book in 1847, and again in 1852. Historians consider it one of China's earliest systematic attempts to understand the West during the nineteenth century.
3. In 1830, the British East India Company's steamship, the *Forbes*, and in 1835, the British firm Jardine, Matheson & Co.'s steamship, the *Jardine*, visited Guangzhou. In the latter incident, Guangzhou civil and military

officials boarded and inspected the ship but prohibited it from entering the port. The incident was reported by *The Chinese Repository*, an English language missionary newspaper published in Macao. *The Chinese Repository* IV (May 1835–Apr 1836), p. 438.

4. Guo Shila (Karl Gützlaff), *Maoyi tongzhi*, quoted in Xiong Yuezhi, *Xixue dongjian yu wan Qing shehui* (Beijing: Zhongguo renmin daxue chubanshe, 2011), p. 114. As to Gützlaff's work in China, see Jessie Gregory Lutz, *Opening China: Karl F. A. Gützlaff and Sino-Western Relations, 1827–1852* (Grand Rapids, MI: William B. Eerdmans Pub. Co., 2008).
5. Wenqing et al. (comp.) *Chouban yiwu shimo*, Daoguang reign (Beiping [Beijing], 1929, reprint of the 1851 edition), 11:18ab. *Chouban yiwu shimo* was a collection of official documents concerning Qing China's foreign affairs. It was first compiled after the death of the Daoguang emperor (r. 1820–1850).
6. Anonymous, *Yifei fanjing wenjian lu* (Beijing: Zhonghua quanguo tushuguan wenxian suowei fuzhi zhongxin, 1995, reprint of the 1857 edition), p. 14.
7. Joseph Needham has traced the invention of the treadmill paddle-mill boat to Zu Chongzhi, a fifth-century official who was interested in mathematicians. Zu built a thousand *li* boat that could cruise several hundred *li* in a day without the help of wind. Late eighth-century records have clear descriptions of a kind of naval boat with two wheels that were attached to its sides and propelled by treadmills. The Southern Song navy also used the same type of boats to fight the Jurchens on the Yangzi River. Treadmill paddle-wheeled vessels were still being used in the Pearl River in the 1920s. Joseph Needham, *Science and Civilisation in China*, Vol. 4 Physics and Physical Technology, Part 2: Mechanical Engineering (Cambridge: Cambridge University Press, 1965), pp. 413–435; Jung-pang Lo, "China's Paddle-wheel Boats: Mechanized Craft Used in the Opium Wars and Their Historical Background," *The Tsing Hua Journal of Chinese Studies* 2:1 (1960), pp. 189–206.
8. Wei Yuan, *Haiguo tuzhi*, 84:27a–28b.
9. Wei Yuan, *Haiguo tuzhi*, 79:16a.
10. J.E. Bingham, *Narrative of the Expedition to China from the Commencement of the War to Its Termination in 1842*, vol. 1 (London: Henry Colburn, 1842), p. 181.
11. Wei Yuan, *Haiguo tuzhi*, 86:2a; Joseph Needham, *Science and Civilisation in China*, vol. 4, part 2, pp. 429–430.
12. Wei Yuan, *Haiguo tuzhi*, 84:27ab.
13. Wei Yuan, *Haiguo tuzhi*, 84:28b–29a; Joseph Needham, *Science and Civilisation in China*, vol. 4 part 2, p. 430.

14. William Bernard, *Narrative of the Voyages and Services of the Nemesis, from 1840 to 1843*, vol. 2 (London: Henry Colburn, 1844), p. 226.
15. Wei Yuan, *Haiguo tuzhi*, 84:16b; Gideon Chen [Chen Qitian], *Lin Tse-hsü: Pioneer Promoter of the Adoption of Western Means of Maritime Defense in China* (New York: Paragon Book Gallery, 1961, reprint of the 1934 Yenching University edition), p. 35.
16. Cao Cheng, *Yihuan baichang ji* (Taipei: Wenhai, 1968, reprint of the 1842 edition), pp. 85–86. Cao Cheng was a gentry man of Zhejiang province.
17. William Bernard, *Narrative of the Voyages and Services of the Nemesis, from 1840 to 1843*, vol. 2, p. 396.
18. J.F Davis, *China during the War and since the Peace*, vol. 1 (London: Longman, Brown, Green, and Longmans, 1852), p. 258.
19. Wenqing (compl.), *Chouban yiwu shimo*, Daoguang reign, 59:48a–49a.
20. Qing officials were also making an effort to improve their warships. By 1840, Lin Zexu built some vessels after the European model. He also purchased an American merchant sailing ship, the *Cambridge*, and armed it with cannons. In 1841, Yi Changhua, prefect of Guangzhou, built a war junk after the Western model. Pan Shicheng, a Guangzhou merchant, contributed money to building four ships after the Western model. For more details, see Gideon Chen, *Lin Tse-hsü*, pp. 32–37.
21. Wenqing (compl.), *Chouban yiwu shimo*, Daoguang reign, 63:16ab.
22. Wei Yuan (ed.), *Haiguo tuzhi*, 89:12a–13b; Wenqing, *Chouban yiwu shimo*, Daoguang reign, 63:15ab; Jung-pang Lo, "China's Paddle-wheel Boats: Mechanized Craft Used in the Opium Wars and Their Historical Background," pp. 189–212.
23. Wenqing, *Chouban yiwu shimo*, Daoguang reign, 63:17ab.
24. George Smith, *A Narrative of An Exploratory Visit to Each of the Consular Cities of China and to the Islands of Hong Kong and Chusan, in Behalf of the Church Missionary Society, in the Years 1844, 1845, 1846* (London: Seeley, Burnside & Seeley, Fleet Street; Hatchard & Son, Piccadilly; J. Nisbet and Co., Berners Street, 1847, second edition), pp. 109–110.
25. *The Chinese Repository* XVI (1847), 104; Charles Gutzlaff, *The Life of Taou Kwang, Late Emperor of China. With Memoirs of the Court of Peking; including A Sketch of The Principal Events in The History of The Chinese Empire during The Last Fifty Years* (London: Smith, Elder and Co., 1852), pp. 221–222.
26. The Chinese man also produced two volumes of diagrams and maps of the stars, asking Smith's opinion about them and asked him to send him some books of Western astronomy. George Smith, *A Narrative of An Exploratory Visit to Each of the Consular Cities of China and to the Islands of Hong Kong*

and Chusan, in Behalf of the Church Missionary Society, in the Years 1844, 1845, 1846, pp. 109–110.
27. I am extremely grateful to Dr. David Gwyn, who generously showed me his own manuscript. In it, Gwyn compares Ding Gongcheng's model engine with various early locomotives designs, thus making my own research possible.
28. Ding Gongchen, *Yanpao tushuo jiyao*, 4:16b.
29. Weqing et al., *Chouban yiwu shim*, Daoguang reign, 63:15ab.
30. For more information about Zheng Fuguang, see Wang Ermin, "Zheng Fuguang yu taixi ke ji zhishi," the Institute of Modern History, Academia Sinica (ed.), *Jindai Zhongguo lishi renwu lunwen ji* (Taipei: The Institute of Modern History, Academia Sinica, 1993), pp. 739–770.
31. Zheng Fuguang, "Huo lunchuan tushuo," in *idem.*, *Jingjing lingchi* (Shanghai: Shangwu, 1936, reprint of the 1847 edition) 3:38a–46a.
32. Zheng Fuguang, "Dingbu huolunchuan jiju tushuo zhiwu," *Jingjing lingshi*, 3:45a–46a.
33. Liang Tingnan, "Lanlun oushuo," *Haiguo sishuo* (Beijing: Zhonghua, 1993), pp. 160–161.
34. Pan Shicheng was one of the *Hong* merchants responsible for dealing with foreign merchants on behalf of the Qing government. Pan worked within the Canton (Guangzhou) system, by which the Qing government confined foreign trade in Guangzhou and appointed only a small numbers of Chinese merchant permitted to trade with foreigners. For the history of the Canton System and the *Hong* merchants, see Liang Jiabin, *Guangdong shisan hang kao* (Taizhong: Donghai daxue chubanshe, 1960); Weng Eang Cheong, *Hong Merchants of Canton: Chinese Merchants in Sino-Western Trade, 1684–1798* (London: Curzon, 1997); Kuo-Tung Anthony Chen, *The Insolvency of The Chinese Hong Merchants, 1760–1843* (Taipei: The Institute of Economics, Academia Sinica, 1990). For Pan Shicheng's sponsoring of shipbuilding during the First Opium War, see Gideon Chen, *Lin Tse-hsü*, pp. 32–37.
35. Aurelius Harland to William Harland June 24, 1847, quoted in David Wright, "Response to Kent Deng," *History of Technology* 25 (2004), Special Issue: the Global History of the Steam Engine, pp. 173–174.
36. As early as 1846, there were two steamboats shuttling between Hong Kong and Guangzhou. In 1848, a company called Hongkong Canton Steam Packet & Co. was formed. This firm ferried business between Hong Kong and Guangzhou. For more information, see Nie Baozhang (ed.), *Zhongguo jindai hangyun shi ziliao*, vol. 1 (Shanghai: Shanghai Renmin, 1983), p. 7. Additionally, P&O provided river service between Hong Kong and Guangzhou; in 1850, P&O established the Hong Kong-Shanghai line.

37. Xin Yuanou, *Zhongguo jindai chuanbo gongye shi* (Shanghai: Shanghai guji chubanshe, 1999), pp. 21–39.
38. *The North China Daily News*, April 21, 1864, quoted in Xin Yuanou, *Zongguo jindai chuanbo gongye shi*, p. 34.
39. Xin Yuanou, *Zhongguo jindai chuanbo gongye shi*, pp. 34–35
40. *The North China Herald*, April 5, 1881.
41. Xin Yuanou, *Zhongguo jindai chuanbo gongye shi*, pp. 35–36.
42. Xin Yuanou, *Zhongguo jindai chuanbo gongye shi*, pp. 40–42.
43. For promoting Western medical knowledge to Chinese people, Hobson published a book on physiology, titled *Quanti xinlun* (*New Treatise on Physiology*), in 1851. For more details, see Chan Man Sing, "Sinicizing Western Science: The Case of *Quanti xinlun*," *T'ung Pao*, 98 (2012), pp. 528–556.
44. He Xin (Benjamin Hobson), *Bowu xinbain*, 1:1a–3a (the 1855 Shanghai edition kept in the Kuo Ting-yee Library, Academia Sinica). The term *diqi* is intriguing; in the classic *The Book of Rite* (*Liji*)'s "Monthly Commandments" (*Yueling*) chapter, the *qi* of heaven (*tianqi*) and the *qi* of earth (*diqi*) are considered the cause of seasonal changes. For example, in the first month of spring, "the *qi* of heaven descends and the *qi* of the earth ascends. Heaven and earth are in harmony. All plants bud and grow." In the third month of spring, "In this month the *qi* of life (*shengqi*) are fully developed; and the positive *qi* (*yangqi*) diffuse themselves. The crooked shoots are all put forth, and the buds are unfolded. Things do not admit of being restrained." However, because of completely different phrasing, the *Bowu xinbian* does not give its readers any impression that its *diqi* had anything to do with the *diqi* in the *Book of Rites*, although nineteenth century learned men could certainly have made an association with the traditional meaning. Because the term was not used later to denote the atmosphere, in the end, it did not make much impact on Chinese understanding of the atmosphere.
45. The chapter also states that the atmosphere consists of 21 per cent oxygen (i.e., *yangqi*, which literally means "life-supporting gas") and 79 per cent hydrogen (i.e., *qingqi*, which literally means "light gas"). It also contains nitrogen (i.e., *danqi*, which literally means "odorless gas"), and carbon dioxide (i.e., *tanqi*, which literally means "carbon gas"). It was the first time such an idea was introduced to the Chinese. See Du Shiran, Lin Qingyuan, and Guo Jinbin, *Yangwu yundong yu Zhongguo jindai keji* (Shenyang: Liaoning jiaoyu chubanshe, 1991), pp. 245–46; David Wright, *Translating Science: The Transmission of Western Chemistry into Late Imperial China, 1840–1900* (Brill: Leiden, 2000), p. 226.

46. After the general discussion of the atmosphere, the *Bowu xinbian* gives two short sections on the barometer and the thermometer. Then it discusses oxygen, hydrogen, nitrogen, and carbon oxide. See *Bowu xinbian*, 1:3a–5b.
47. To translate the term for atmosphere he essay uses the classical term *yinyun*, which is from the Chinese classic *Yijing* (*Book of Change*). *Yijing* uses the same term to denote the condition created by the mixture of the *yin* and the *yang* at the beginning of the universe.
48. No document describes the facilities and the skills of the Anqing Arsenal. Sun Yutang (ed.), *Zhongguo jindai gongyeshi ziliao diyi ji*, pp. 251–252.
49. According to the *North China Herald*, "the boiler was made of a compound resembling zinc; the diameter of the cylinder was one inch and seven tenths, and the speed attained by the engine was 240 revolutions per minute." *The North China Herald*, 5th September 1868.
50. Zeng Guofan, *Zeng Wenzhang gong quanji: qiushizhai riji leichao* (Taipei: Wenhai, 1974, reprint of the 1870s edition), 1:55ab, translation based on Gideon Chen, *Tseng Kuo-fan: Pioneer Promoter of the Steamship in China* (New York, 1968, reprint of 1935 edition), p. 40.
51. The historian Chen Qitian argues that Zeng's "two of the front holes" might have referred to two valves. When they were closed, the steam goes into "the other front hole" or front part of the cylinder. Gideon Chen, *Tseng Kuo-fan*, pp. 40–41.
52. John Fryer continues: "The steamer [the *Huanghu*], which was of twenty-five tons measurement, was able at her trial trip on the Yangtze in the year 1865 to make 255 *li* or about 85 miles in fourteen hours, and to do the return journey in less than eight hours." *North China Herald*, 29th January, 1880.
53. *The North China Herald*, 5th September 1868; Xin Yuanou, *Zhongguo jindai chuanbo gongye shi*, p. 94.
54. *The North China Herald*, 5th September 1868.
55. *Yangwu yundong*, vol. 4, p. 16.
56. On 28th January 1864, Zeng Guofan saw the steamboat built by his subordinate Cai Guoxiang: "I went out of town and down to the river to see a small steamboat built by Cai Guoxiang, two *zhang* and eight or nine *chi* long [between 8.4 and 9.45 feet]. Since on reaching the middle of the river we had traveled [only] eight or nine *li* [2.86 or 3.22 miles], I calculate that in two hours we would have traveled around 25 or 26 *li* [around 8.95 or 9.3 miles]. Larger versions of this will be built after this model, and more will be built." Cai Guoxiang was a commander of the *Xiang* Army, but we know nothing else about him. Historians attribute this steamboat to Xu Shuo and Hua Hengfang, arguing that it was Cai Guoxiang who supervised the building work because Zeng Guofan put Cai's name in his diary.

However, there is no evidence to support such a speculation. *The North China Herald* only recorded one trial steamboat, which was the one built by Xu and Hua in 1863 one and the account matches John Fryer's reminiscence of 1880. *The North China Herald*, 5th September 1868 and 29th January, 1880. See also Gideon Chen, *Tseng Kuo-fan*, p. 41; David Wright, "Careers in Western Science in Nineteenth-Century China: Xu Shou and Xu Jianjin," *Journal of the Royal Asiatic Society* 5:1 (1995), p. 64; Bai Guangmei and Yang Gen, "Xu Shou yu Huanghu hao lunchuan,' in Yang Gen (ed.), *Xu Shou he Zhongguo jindai huaxueshi* (Beijing: Kexue jishu wenxian chubanshe, 1986), pp. 161–162.

57. David Wright, "Careers in Western Science in Nineteenth-Century China: Xu Shou and Xu Jianjin," p. 64.
58. *The North China Herald*, 5th September, 1868.
59. The Chinese scholar Song Ziliang points out that it is impossible to build a steam engine according to the simple description of the *Bowu xinbian*. Song Ziliang, "Anqing nei junxiesuo he Zhongguo diyisao zhengqichuan," *Chuanshi yanjiu* No.6 (1993), p. 2.
60. According to the Chinese historian Xin Yuanou, the steel parts of the engine's shaft, boiler, and cylinder were purchased from abroad. Most of the parts were made in the arsenal. Xin Yuanou, *Zhongguo jindai chuanbo gongye shi*, p. 101.
61. Demetrius C. Boulger, *The Life of Sir Halliday Macartney* (London, New York: J. Lane, 1908), p. 79. Halliday Macartney originally served in the British Army in China as a military medical doctor. He later left the British service and joined Li Hongzhang's army in 1863.
62. Unfortunately we do not know the quantity of these machine tools. Demetrius C. Boulger, *The Life of Sir Halliday Macartney*, pp. 125–132.
63. In April 1863 Li Hongzhang told Zeng Guofan that he was trying to hire foreign technicians and purchase machine tools (or tools that made guns) from Hong Kong. Li Hongzhang, *Li Wenzhong gong quanji*, Letters to Friends and Colleagues (Jinling [Nanjing]: 1908), 3:16b.
64. Zhongguo jindai bingqi gongye dang'an shiliao bianweihui (ed.), *Zongguo jindai bingqi gongye dang'an shiliao*, vol. 1 (Beijing: Bingqi gongye chubanshe, 1993), pp. 50–51.
65. Sun Yutang (ed.), *Zhongguo jindai gongyeshi ziliao*, vol. 1, p. 253
66. Sun Yutang (ed.), *Zhongguo jindai gongyeshi ziliao*, vol. 1, p. 260.
67. Yung Wing, *My Life in China and America* (New York, 1978, reprint of the 1909 edition), 149–151. Rong Hong was one of the earliest Chinese who received Western education. He was educated in Macao and Hong Kong 1830s and 1840s as a child in missionary schools. He went to the United Stated in 1847 and graduated from Yale College in 1854. Then he went back to China.

References

Anonymous, *Yifei fanjing wenjian lu* 夷匪犯境聞見錄 (Beijing: Zhonghua quanguo tushuguan wenxian suowei fuzhi zhongxin, 1995, reprint of 1857 edition).

Bai Guangmei 白广美 and Yang Gen 杨根, "Xu Shou yu Huanghu hao lunchuan 徐寿与黄鹄号轮船" in Yang Gen (ed.), *Xu Shou yu Zhongguo jindai huaxue shi* 徐寿与中国近代化学史 (Beijing: Kexue jishu wenxian chubanshe, 1986), pp. 142–153.

Bernard, William, *Narrative of the Voyages and Services of the Nemesis, from 1840 to 1843*, 2 volumes (London: Henry Colburn, 1844).

Bingham, J.E., *Narrative of the Expedition to China from the Commencement of the War to Its Termination in 1842*, vol. 1 (London: Henry Colburn, 1842).

Boulger, Demetrius Charles, *The Life of Sir Halliday Macartney* (London: Bodley Head, 1908).

Cao Cheng 曹晟, *Yihuan baichang ji* 夷患備嚐記 (Taipei: Wenhai, 1968, reprint of the 1842 edition).

Chan Man Sing, 'Sinicizing Western Science: The Case of *Quanti xinlun*', *T'oung Pao* 98 (2012), pp. 528–556.

Chen, Gideon (Chen Qitian 陳其田), *Lin Tse-hsü: Pioneer Promoter of the Adoption of Western Means of Maritime Defense in China* (New York: Paragon Book Gallery, 1961, reprint of the 1934 Yenching University edition).

Chen, Gideon (Chen Qitian 陳其田), *Tseng Kuo-fan; Pioneer Promoter of the Steamship in China* (New York: Paragon Book Gallery, 1961, reprint of 1935 edition).

Chen, Kuo-Tung Anthony, *The Insolvency of The Chinese Hong Merchants, 1760–1843* (Taipei: The Institute of Economics, Academia Sinica, 1990).

Davis, J.F., *China during the War and since the Peace*, vol. 1 (London: Longman, Brown, Green, and Longmans, 1852).

Ding Gongchen 丁拱辰, *Yanpao tushuo jiyao* 演礮圖說輯要 (1945 reprint of the 1843–1851 edition, kept in the Xiamen University Library).

Du Shiran 杜石然, Lin Qingyuan 林庆元, and Guo Jinbin 郭金彬, *Yangwu yundong yu Zhongguo jindai keji* 洋务运动与中国近代科技 (Shenyang: Liaoning jiaoyu chubanshe, 1991).

Gutzlaff, Charles, *The Life of Taou Kwang, Late Emperor of China. With Memoirs of the Court of Peking; including A Sketch of The Principal Events in The History of The Chinese Empire during The Last Fifty Years* (London: Smith, Elder and Co., 1852).

He Xin 合信 (Benjamin Hobson), *Bowu xinbian* 博物新編, the 1855 Shanghai edition kept in the Kuo Ting-yee Library, Academia Sinica.

Li Hongzhang 李鴻章, *Li Wenzhong gong quanji* 李文忠公全集 (Jinling: 1908 edition).

Liang Jiabin 梁嘉彬, *Guangdong shisanhang kao* 廣東十三行考 (Shanghai: Shanghai shangwu, 1937).

Liang Tingnan 梁廷楠, *Haiguo sishuo* 海國四說 (Beijing: Zhonghua, 1993).

Lo Jung-pang 羅榮邦, "China's Paddle-wheel Boats: War and Their Historical Background", *Tsing Hua Journal of Chinese Studies* 清華學報, 2:1 (1960), pp. 189–212.

Lutz, Jessie Gregory, *Opening China: Karl F. A. Gützlaff and Sino-Western Relations, 1827–1852* (Grand Rapids, MI: William B. Eerdmans Publishing, 2008).

Needham, Joseph, *Science and Civilisation in China*, Vol. 4 Physics and Physical Technology, Part 2: Mechanical Engineering (Cambridge: Cambridge University Press, 1965).

Nie Baozhang 聶宝璋, *Zhongguo jindai hangyun shi ziliao* 中国近代航运史资料, 2 volumes (Shanghai: Shanghai Renmin 1983).

Smith, George, *A Narrative of An Exploratory Visit to Each of the Consular Cities of China and to the Islands of Hong Kong and Chusan, in Behalf of the Church Missionary Society, in the Years 1844, 1845, 1846* (London: Seeley, Burnside & Seeley, Fleet Street; Hatchard & Son, Piccadilly; J. Nisbet and Co., Berners Street, 1847, second edition)

Song Ziliang 宋子良, "Anqing nei junxiesuo he Zhongguo diyisao zhengqichuan 安庆内军械所和中国第一艘蒸汽船," *Chuanshi yanjiu* 船史研究 No.6 (1993), pp. 1–7.

Sun Yutang 孫毓棠, *Zhongguo jindai gongyehsi ziliao diyi ji: 1840–1895* 中國近代工業史資料第一輯, 2 volumes (Beijing: Kexue chubanshe, 1957).

The Chinese Repository.

The North China Herald.

Wang Dahai 王大海, *Haidao yizhi* 海島逸志 (the 1843 reprint of 1791 edition, kept in the Fu Ssu-nien Library, Academia Sinica, Taipei).

Wang Ermin 王爾敏, "Zheng Fuguang yu taixi ke ji zhishi 鄭復光與泰西科技知識," The Institute of Modern History, Academia Sinica (ed.), *Jindai Zhongguo lishi renwu lunwen ji* 近代中國歷史人物論文集 (Taipei: The Institute of Modern History, Academia Sinica, 1993), pp. 739–770.

Wei Yuan 魏源 (comp.), *Haiguo tuzhi*, 海國圖志 in the *Xuxiu Xikuquanshu* 續修四庫全書 series, vol. 744–745 (Shanghai: Shanghai guji chubanshe, 1995–2002, reprint of the 1876 edition).

Weng, Eang Cheong, *Hong Merchants of Canton: Chinese Merchants in Sino-Western Trade, 1684–1798* (London: Curzon, 1997).

Wenqing 文慶 et al. (comp.), *Chouban yiwu shimo* 籌辦夷務始末, *Daoguang* reign 道光朝 (Beiping: National Palace Museum, 1930, reprint of 1856 edition).

Wright, David, "Careers in Western Science in Nineteenth-Century China: Xu Shou and Xu Jianjin," *Journal of the Royal Asiatic Society* 5:1 (1995), pp. 49–90.

Wright, David, "Response to Kent Deng," *History of Technology* 25 (2004), Special Issue: the Global History of the Steam Engine, pp. 173–176.

Wright, David, *Translating Science: The Transformation of Western Chemistry into Late Imperial China, 1840–1900* (Leiden: Brill, 2000).

Xin Yuanou 辛元欧, *Zhongguo jindai chuanbo gongye shi* 中国近代船舶工业史 (Shanghai: Shanghai guji chubanshe, 1999).

Xiong Yuezhi 熊月之, *Xixue dongjian yu wan Qing shehui* 西学东渐与晚清社会, revised edition (Beijing: Zhongguo renmin daxue chubanshe, 2011).

Yung Wing, *My Life in China and America* (New York, 1978, reprint of 1909 edition).

Zeng Guofan 曾國潘, *Zeng Wenzheng gong quanji* 曾文正公全集 (Taipei: Wenhai, 1974, reprint of the 1876 edition).

Zheng Fuguang 鄭復光, *Jingjing lingchi* 鏡鏡詅癡 (Shanghai: Shangwu, 1936, reprint of the 1847 edition).

Zhongguo jindai bingqi gongye dang'an shiliao weiyuanhui 中國近代兵器工業檔案史料委員會 (comp.), *Zhongguo jindai bingqi gongye dang'an shiliao* 中國近代兵器工業檔案史料 (Beijing: Bingqi gongye chubanshe, 1993).

Zhongguo shixuehui 中國史學會 (ed.), *Yangwu yundong* 洋務運動, 8 volumes (Shanghai: Shanghai renmin chubanshe, 1961).

CHAPTER 3

Translating Heat: Tackling Old Conceptions with New Ideas, 1855–1868

Chapter 2 briefly discussed texts such as the *Bowu xinbian* written by foreign missionaries that introduced the basic scientific ideas of vacuum and atmospheric pressure to Chinese readers. To some extent, those texts made the trials of China's first domestically built steamer, the *Huanghu*, possible. Although engine building involved engineering technicalities including machine tooling, metal smelting and casting, and gearing, there was no denying that the science of heat was the heart of steam engine technology.

This chapter explores how the basic nineteenth-century understanding of heat was introduced to China through texts composed by foreign missionaries from the 1850s to the 1860s. The problem was not only inventing new terms for new ideas, such as temperature, but also redefining conventional terms, especially "heat" and "fire," which had already been embedded in Chinese language and natural philosophy for centuries. Tackling old conceptions meant dissecting them through both Western and Chinese linguistic and philosophical examinations. On the basis of redefined terms, new concepts of temperature, thermal expansion and contraction, the effect of heat on the state of matter, and heat diffusion could be phrased in Chinese, allowing Chinese readers to approach thermal phenomena and the steam engine through a new set of concepts that Ding Gongcheng and Zheng Fuguang did not have. Such an intellectual enterprise could only be done through collaboration between Chinese and foreign translators.[1]

QI, *YIN-YANG* AND *WU XING*

By the time missionary publications appeared in the treaty ports, the Chinese literati were interested in natural studies by following the centuries-old Confucian tradition of *gewu* (the investigation of things) and *qiongli* (the exhaustion of principles).[2] They would have learned from their daily lives that friction creates heat, hot air rises but cold air sinks, combustion generates heat and light, and living humans and animals are warm but dead ones are cold. They would also have recognized different levels of heat or cold, although they did not have a term for temperature, which required the invention of the thermometer. Linguistically, they would have used the word *re* to describe the hot condition or to denote the act of heating up something. In contrast, the word *leng* (cold) was used to describe the cold condition. Both words were mainly used as adjectives or verbs. Conceptually, their explanation of thermal phenomena would have employed *qi*, *yin-yang*, and five phases (*wu xing*). This theory is too complex to be related here but, put simply, it argues that there was *qi* at the beginning of the universe. *Qi* can be divided into two types, positive (*yang*) and negative (*yin*) *qi*. Different levels of positive and negative *qi* formed five phases: metal, wood, water, fire, and earth. The mutual-production-and-mutual-conquest relationship between the five phases created myriad things.[3] In this understanding of cosmology, *qi* has multiple meanings. It could mean, as Nathan Sivin argues, the material stuff in everything but also the vital energy that makes things grow and develop.[4] *Qi* could also mean air, steam, or breath. Through that set of conceptions and terminology, one might explain combustion itself as fire, so an object is hot because its nature belongs to the category of *yang* or fire.[5] A good example can be found in the writings of Song Yingxing, a low-ranking Ming dynasty scholar-official who traveled around China to collect information about technology and wrote the famous *Tiangong kaiwu* (*The Exploitation of the Works of Nature*, first printed in 1637). He argued that solid objects and *qi* are interconvertible, and charcoal and coal are actually solidified fire. According to Christopher Cullen, Song could have measured the conversion ratio of weight before and after wood is burned.[6] Seventeenth-century European alchemists, who built their theoretical foundation on the four elements of Aristotelian natural philosophy, approached the problem of combustion in a similar way. They postulated that substances can burn because they are rich in phlogiston, which has no or even negative weight. Heated metal yields calx, and combustion releases phlogiston into the air, which becomes phlogisticated and stops supporting animal life but helps plant growth. Such an idea was embraced by Western chemists until the late eighteenth century.[7]

In the nineteenth century, neither *qi* nor phlogiston was sufficed for understanding the principles of the steam engine. The core of the problem was more than understanding that steam pressure moved the piston or metaphysically imagining that steam contains energy. Rather, the issue was realizing that heat and motive force are two aspects of energy and their relationship is calculable. This insight required new theory to be translated into Chinese.

HEAT IS A SUBSTANCE, BUT NOT *QI*: THE *BOWU XINBIAN*

From the mid-1850s, foreign missionaries such as the British physician Benjamin Hobson had started to publish popularized science texts as a means of proselytization in China. Hobson was trained at University College London as a physician before he was sent by the London Missionary Society to the Portuguese settlement of Macao in 1839. In 1843, he started to practice in the British colony of Hong Kong and later, in 1848, set up a clinic in Guangzhou, where he learned Chinese. He soon started to work with his Chinese teacher on the translation of a treatise on physiology, the *Quanti xinlun* (*New Treatise on Physiology*, 1851), and then a book on natural philosophy, the *Bowu xinbian* (*Natural Philosophy*). He published the latter in Guangzhou and later in Shanghai in 1855.

Hobson wrote the *Bowu xinbian* on the basis of the *Elements of Physics or Natural Philosophy*, a textbook written in nontechnical language for medical and general students by the British physician Neil Arnott (1788–1874).[8] It contains five parts: Part 1 discusses the constitution of materials; Part 2 examines mechanics; Part 3 focuses on hydrodynamics; Part 4 explores imponderable substances, which are heat, light, electricity, and magnetism; and Part 5 discusses astronomy. It was first published in 1827, and from the third edition in 1829 on, it was expanded into a two-volume book but retained the original structure. The sixth edition was published just before Arnott's death in 1874.[9] Unfortunately, we do not know which edition Hobson consulted. He could have read the fifth (1833) edition as a medical student in London, and hence, we shall use that edition as the point of comparison.

The structure of the *Bowu xinbian* does not resemble that of *Elements of Physics*. Hobson's method of translation was to "select the essentials and translate them into Chinese."[10] Such a method involved close cooperation with his Chinese collaborators, who were well-versed in Chinese classics and philosophy. Historians have often quoted the account of John Fryer, a British missionary who from 1867 worked as a translator at the Qing government's Jiangnan Arsenal translation department, to describe the translation process:

> The foreign translator, having first mastered his subject, sits down with the Chinese writer and dictates to him sentence by sentence, consulting with him whenever a difficulty arises as to the way the ideas ought to be expressed in Chinese, or explaining to him any point that happens to be beyond his comprehension. The manuscript is then revised by the Chinese writer, and any errors in style, &c., are corrected by him. In a few cases the translations have been carefully gone over again with the foreign translator, but in most instances such an amount of trouble has been avoided by the native writers, who, as a rule, are able to detect errors of importance themselves, and who, it must be acknowledged, take great pain to make their style as clear and the information as accurate as possible.[11]

A similar process would have applied to Hobson and his Chinese collaborator. Hobson would have had to verbally explain the contents of the book in colloquial Chinese to his Chinese collaborators, who composed the texts in classical Chinese according to their understanding. They would have to discuss the meanings of the terms and the style of composition with each other. As Chan Man Sing, Law Yuen Mei, and Kwong Wing Hang have argued, hybridization was unavoidable because of the linguistic and intellectual influence of the Chinese translators.[12] Indeed, David Wright has pointed out that Fryer's account underappreciates his Chinese colleagues' contribution in the shaping of the contents of translation because their linguistic and philosophical capacity to transform the foreign translators' rough ideas into readable and publishable texts was essential. Without their Chinese collaborators, as Wang Yangzhong and Iwo Amelung have shown, even the most competent foreign translators who commanded a good understanding of Chinese could render clumsy, difficult-to-read texts.[13] Thus, Hobson and his Chinese collaborator had to redefine the old terms "fire" and "*qi*" to build a foundation on which they could explain heat.

The content about heat in Arnott's *Elements of Physics* follows the early nineteenth-century mainstream theory that considers heat an all-pervasive, imponderable fluid called "caloric." The book argues that heat as the cause of winter and summer is essential to life. Measurable by a thermometer, heat diffuses among neighboring bodies until all reach the same level of heat. Heat spreads through conduction and convection. It may also radiate. The flow of heat in and out of a body causes it to expand or contract. Heat can also cause changes in state because heat overcomes the adhesive force that binds particles in a body. In the process of changing state, more heat may be needed but could be "hidden" from the thermometer. This hidden heat is called "latent heat," while lack of heat is "cold." Different substances have different capacities to absorb heat before they change their states, and the amount of heat that is required to raise the temperature by one degree is "specific heat." The sun, electricity, combustion, chemical reactions, friction, and life are the sources of heat.

Because heat is the core of understanding the steam engine, some explanation here, at the risk of oversimplification, is justified. The concept of the caloric was invented by the French chemist Antoine Lavoisier (1743–1794) as one of the consequences of discovering oxygen in the late eighteenth century.[14] Arnott's explanation that temperature is the measurement of heat is a caloric view. It was known to the calorists that ice requires more heat to turn into water and water needs more heat to be transformed into steam without changes in temperature; the eighteenth-century Scottish physician and chemist Joseph Black (1728–1799) named this phenomenon "latent heat" and argued that heat spent in melting ice was not destroyed but converted into a different state, and hence could not be detected by the thermometer.[15] That discovery decoupled temperature and heat. Heat capacity, or capacity for heat, is another concern because an object might contain little heat but could be registered at a high temperature by the thermometer. Furthermore, different materials expand at different rates, and even the same material expands at different rates because its heat capacity may change at different temperatures. Gas, which was considered to be close to caloric, absorbs more heat at a higher temperature. Heat capacity can be measured by the thermometer, which is a vessel lined with ice and the quantity of heat given out by any body placed in it is indicated by the quantity of water collected from the melted ice.[16] These issues were important in thermal physics and particularly in understanding the steam engine.

The early steam engine invented by Thomas Newcomen was extremely inefficient because creating a partial vacuum by injecting cold water into the cylinder would lower its temperature and often caused a breakdown. Hence, the engine could only be used in coal mines with coal waste as fuel. To reduce costs, an engineer carefully measured the Newcomen engine's "duty" in pounds of water lifted one foot high by the burning of a bushel of coal and kept constant the temperatures of most parts of the engine. James Watt's engine was a far more radical change. He was fully aware of the issues of latent heat and the loss of heat by cooling the cylinder.[17] To save fuel, he invented the external condenser and a steam jacket to keep the cylinder warm. He also employed the "cut-off" technique, which cut off the supply of steam to the cylinder, allowing steam to expand for the rest of the power stroke and ensuring the maximum output of power from a given amount of steam.[18] From the early 1800s to the 1850s, the progress from the Watt engine to the high-pressure engine as well as the compound engine relied on constant improvement in casting the cylinder and designing the boiler, allowing steam pressure high enough to expand twice in two different cylinders, thereby drastically reducing the consumption of coal.[19] However, there was not yet a general theory to explain the relationship between heat and motive force.

In 1824, the young French military engineer Sadi Carnot (1796–1832) reflected upon the power of the steam engine and its relationship with heat, arguing that, similar to the flow of water that drives the waterwheel, the flow of heat from a hot reservoir (cylinder) to a cold one (condenser) drives the engine. The greater the temperature difference, the more powerful the motive force. He also argued that the working fluid must be allowed to expand on its own while cooling in order to extract the maximum mechanical effect from the heat. Such an idea benefited from Watt's cut-off technique. Most importantly, for an ideal engine, Carnot also argued that the working fluid should ultimately return to its original state (in temperature and pressure). Hence, his ideal heat engine works in a cycle.[20] Carnot's brilliant work was ignored until the 1830s, when the French engineer Émile Clapeyron (1799–1864) reformulated and mathematicized his concept. The essential part of Clapeyron's mathematical reformulation involved employing the mathematical technique of infinitesimal calculus to calculate the relationship between heat and mechanical work.[21] In the 1840s, Clapeyron's works were translated into German and English, slowly but gradually attracting attention from engineers and physicists.[22]

Arnott did not mention anything about Carnot's theory, which was limited to a small circle of elite physicists and engineers by the 1830s. However, like most writers of steam engine technology at that time, Arnott recognized heat's essential role in generating steam, whose expansive properties drive the piston.[23] Hobson would not have known about Carnot's theory either. To compose the *Bowu xinbian*, Hobson selected parts of the contents from Arnott's *Elements* and included some other materials related to the steam engine in his book. The most apparent of these is the diagram of a Watt's engine from an unknown source.

The *Bowu xinbian* has three volumes. The first one has five sections, which are on the atmosphere, heat, water, light, and electricity. The second volume focuses on astronomy, and the third, physiology. Chapter 2

3 TRANSLATING HEAT: TACKLING OLD CONCEPTIONS WITH NEW IDEAS...   63

mentioned that the opening section of the *Bowu xinbian* is the *Diqi lun* (atmosphere), which is followed by a short essay on the air pump. That section also introduces the barometer (*fengyu zhen*, literally "wind and rain pointer") and the mercury thermometer (*hanshu zhen*, literally "cold and heat pointer") (Fig. 3.1). The term "*hanshu zhen*" indicates that the

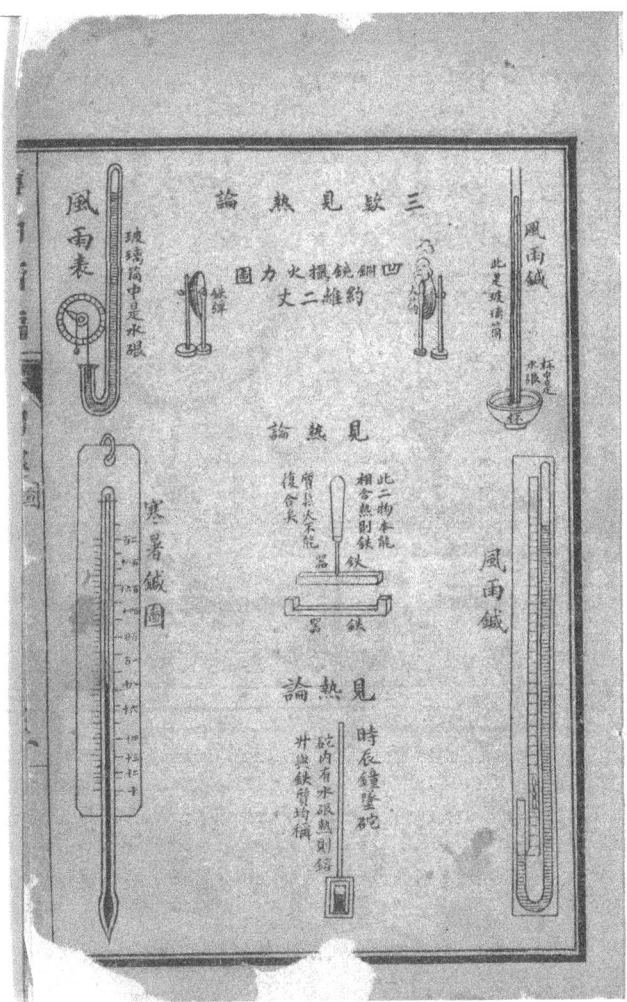

Fig. 3.1 The *Bowu xinbian*'s illustrations of the thermometers. Source: He Xin, *Bowu xinbian* (undated edition kept in the Kuo Ting-yee Library, Academia Sinica, Taipei)

instrument points to the level of cold and hot. However, the term "*hanshu*," which has long been used to describe the coldness and hotness of winter and summer in Chinese literary writings, still implies a separation between cold and heat.[24]

Notably, the text does not invent a new term to translate "temperature." Rather, it explains that because mercury is soft and liquid-like, the mercury in the glass tube will rise up when it touches a hot object but fall down if the object is cold. One can fix a tube of mercury on a board and draw a scale next to it. When the river is frozen, the mercury reaches 32 degrees on the scale, but the number gets higher when it is hotter. The human body is normally 96 degrees, so if one reaches 120 degrees, this indicates a fever. Boiling water is 212 degrees and combustion can reach 1000 degrees.[25] Such a numerical description of the level of hotness and coldness would have been new to Chinese readers in the 1850s. Although the text does not explain why the thermometer functions in this way, it does imply that mercury expands when heated but contracts when it becomes cold.

However, the thermometer is not just an instrument capable of giving readings of temperature; it raises theoretical and phenomenological questions. Hobson and his Chinese collaborator had to reshape Arnott's caloric argument for Chinese readers. The section "On Heat" (*Re lun*) starts with the statement that heat is the most important thing in the world. Everything relies on heat to grow and develop, although it is imponderable (*wuxing wuzhi*). The learned people in the West had classified six sources of heat: the sun, fire (combustion), electricity, live bodies, chemical reactions, and friction. It further explains that heat from the sun travels along with light, and its function is to give life. Fire can generate heat and light at the same time, but heat cannot reach as far as light can. Electricity is generated by the "interaction" (*ganfa*, literally "sense and develop") between the earth and the atmosphere (*qi*), causing lightning that flashes through the sky. Westerners can produce electricity with materials or machines. Living bodies naturally generate heat but cannot emit light. Heat can also be generated by chemical reactions such as pieces of rotten wood being turned into fungi or changes in states of matter. Additionally, heat can be generated by two objects striking against each other such as making fire by using a hand drill or striking two rocks against each other.

Then, the text states that heat is diffusible but indestructible, which is the concept of the conservation of heat, and hence, heat is transmitted from one body to another until the two bodies reach the same level of hot and cold. However, the speed of heat diffusion varies. Heat passes in and out of metal much faster than cloth and wood. This is why people wear cotton or wool clothes rather than metal armor, which cannot keep human bodies warm. The *Bowu xinbian* further explains why there is no such thing as "cold" (*leng*):

> Although [heat] is imponderable, it is still a substance in the universe, and it is certainly different from cold. Cold is like bland taste, but heat strong flavor. Bland taste is [similar to] emptiness, but with flavor it could be [sensed]. Cold is void, with heat it could be [sensed]. Therefore, cold is not a substance but heat is, just as blandness is not a thing but flavor is something.[26]

Hobson's text presents a material understanding of heat. The argument is simple and apparent from a modern perspective but would have posed an intellectual challenge to nineteenth-century Chinese readers because the text followed a paradigm that was completely alien to their own. By focusing on the word *re* (heat), the text shifts the conceptual attention from *yin-yang* and the five phases to heat as the imponderable substance itself. In other words, one has to think of heat (*re*), not *yang* or fire, as the cause of thermal phenomena. Yet, using the analogy of taste to argue that cold is lack of heat is quite awkward because flavor is a human sensation, not a substance.

The text subsequently explains the nature of heat:

> Some might doubt that heat is transformed from the atmosphere (*diqi*). Had it been so, there would be heat wherever there is air (*qi*) [and *vice versa*]. Westerners did an experiment by using an air pump to pump air out of a bottle, and then placing a hot object into it. The heat [of the object] did not increase or decrease, hence we know that heat and air are two different things. [The distinction] is similar to adding flavor to water: water and flavor are two different things. Some suspect that heat is the [metaphysical] *qi* within fire [combustion]. If so, when there is fire, there would be heat. However, why can heat be generated by rubbing hands? Rotten things immersed in water also generate heat. Such phenomena have nothing to do with fire. [The light emitted by] fireflies [or] marine micro-organisms cannot burn anything. Such things do not generate any heat! Hence, we know that heat is not the *qi* within fire. Only the combination of heat and light [combustion] that can burn things can be called fire.[27]

The suggestion that the atmosphere might be the source of heat is intriguing. It is likely that the Chinese person who posed such a question to Hobson was imagining that the tropical climate in Guangzhou was itself a source of heat. The person might also have imagined that heat generated by fire was some kind of *qi*. Such confusion might have something to do with the fact that the word *qi* denoted both air and the metaphysical *qi* before the modern term *kongqi* was widely accepted as the standard term for air after the late nineteenth century. Besides, the mentions of heat generated by friction and decomposition as well as the light emitted by fireflies and marine micro-organisms intends to explain that heat is not *qi*, nor is heat metaphysical fire, although combustion generates heat.

Yet, Hobson's efforts to refute the *qi* theory causes a problem about whether a vacuum can retain the heat of a hot object. A vacuum is an effective heat insulator, but heat radiates through empty space and vacuums, and hence, the temperature of the hot object inside a vacuum would certainly change, provided that the temperature surrounding the vessel is different. Indeed, the 1833 edition of the *Elements of Natural Philosophy* reads:

> The diffusion of heat by radiation, as it takes place in an instant to any distance, and begins whenever there is any inequality of temperature between bodies exposed to each other, would produce instant balance of temperature through out nature, but that heat leaves and enters bodies with readiness depending on their internal conducting powers, and on the condition of their surfaces.[28]

Regardless of that error, the *Bowu xinbian*'s point is that heat has nothing to do with the metaphysical *qi*. Unfortunately, keeping using the same term *qi* to denote air and fire to indicate combustion would certainly cause some difficulty in understanding the rationale behind the argument.

Next, the *Bowu xinbian* discusses the relationship between heat and light. It states that heat and light are two different materials, and humans can see light but only feel heat. The two materials coexist. However, all reflective surfaces can reflect both heat and light. For example, a convex lens can focus both heat and light from the sun to make fire. However, light can penetrate glass, but heat cannot. Unfortunately, the text does not go beyond these simple descriptions.

Latent Heat or "Intrinsic Heat"?

Following the "On Heat" section, the *Bowu xinbian* has a section on changes of state. It explains that "[substances of] all three states [solid, liquid, and gas] contain 'intrinsic heat' (*benre*). Gases contain the most of it, fluids next, and solids the least. It is nature."[29] Then it states that additional heat is needed to transform solids to liquids and liquids to gases: "everything in the world contains intrinsic heat and living things rely on things to live and reproduce."

The description of "intrinsic heat" is curious. When the text explains that more heat is needed to change state, it reads like a description of latent heat. However, there is no such thing as "intrinsic heat."[30] This error could have been the result of a misunderstanding between Hobson and his Chinese collaborator. Unfortunately, we do not have the draft manuscript of the *Bowu xinbian*, so we cannot judge who made the mistake.

Therefore, the *Bowu xinbian* provides rudimentary knowledge about temperature, the nature of heat, heat diffusion and the effect of heat on changes of state, although it does not explain the effects of heat on volume, and its description of laten heat and the effort to distinguish heat and cold is problematic. On this basis it explains why heat is not *qi* or the metaphysical fire. Such way of explanation is a departure from the traditional natural philosophy that classifies natural phenomena by the categories of *yin-yang* and the five phases of *qi*.

More Argument Against *Qi*: The *Xia'er Guanzhen*

Around the same time, a slightly more logical explanation of heat was published in three short essays in the missionary periodical *Xia'er guanzhen* (*Chinese Serial*) in 1855. The anonymous author (or authors) might have had a missionary background because the author stated that "if one is willing to understand what heat is, [one has to realize that] it is the mystery of God's creation and cannot be understood by common-sense."[31]

The first essay, which appeared in issue eight of the *Xia'er guanzhen* and was entitled "A General Discussion on Heat (*Reqi zhi zonglun*)," gives an overview of the nineteenth-century conception of heat. Notably, the essay uses "*re*" and "*reqi*" interchangeably. Such usage is commonly seen in Chinese writings. However, the term "*reqi*" literally suggests hot air, not just heat. This confusion shows that the author or authors did not seem to be careful about the terminological distinction between heat (*re*) and air (*qi*).

The essay states that heat changes the state of matter, moves machines, causes seasonal changes, and drives the fire-wheel engine.[32] Furthermore, it asserts that one should distinguish heat from human sensations of hot or cold because all bodies contain heat and human sensations cannot be trusted as the basis of measuring the level of heat. Rather, heat can only be measured by observing how much it changes the states of matter. Additionally, the speed of heat flow depends on the properties of the bodies. For example, metal transmits heat much faster than wood. Hence, there is no such thing as "cold" because this is lack of heat. Even ice and snow, which people consider to be extremely cold, contain heat and would melt frozen mercury.[33]

The essay goes on to provide a short description of two simple experiments presenting the concept of thermal expansion and contraction. In the first experiment, one fills a glass tube with colored water and observes the rise of the water level when the tube is heated. The other experiment was to take two iron blocks, one rectangular and the other square-U shaped, which fit perfectly together at room temperature, then heat up one of the blocks and observe how they cease to fit together (Fig. 3.2).

The second essay was published in issue nine (1855) of the *Xia'er guanzhen*. It gives a short description of how to produce the mercury thermometer and outlines the idea of the two reference points of temperature: the freezing and boiling points of water, which are 32 and 212 degrees Fahrenheit, respectively. However, the essay does not provide a translation of the term "temperature," which had no equivalent in the Chinese language at the time.

The third essay in the *Xia'er guanzhen*'s issue ten discusses thermal expansion. It gives a list of the expansion rates of different materials including water, lead, silver, copper, gold, and iron. It also tells readers that thermal expansion and contraction is an issue that technicians should consider in cases such as bridge building. Furthermore, the expansion rates of liquids and gases are larger than solids.

Like the *Bowu xinbin*, the *Xia'er guanzhen* essays show the fundamental conceptual difference between the Western material understanding of heat and the Chinese conceptions of *yin-yang* and the five phases of *qi*. To make a clear distinction, the third essay has a passage discussing the question of *qi* and fire:

3 TRANSLATING HEAT: TACKLING OLD CONCEPTIONS WITH NEW IDEAS... 69

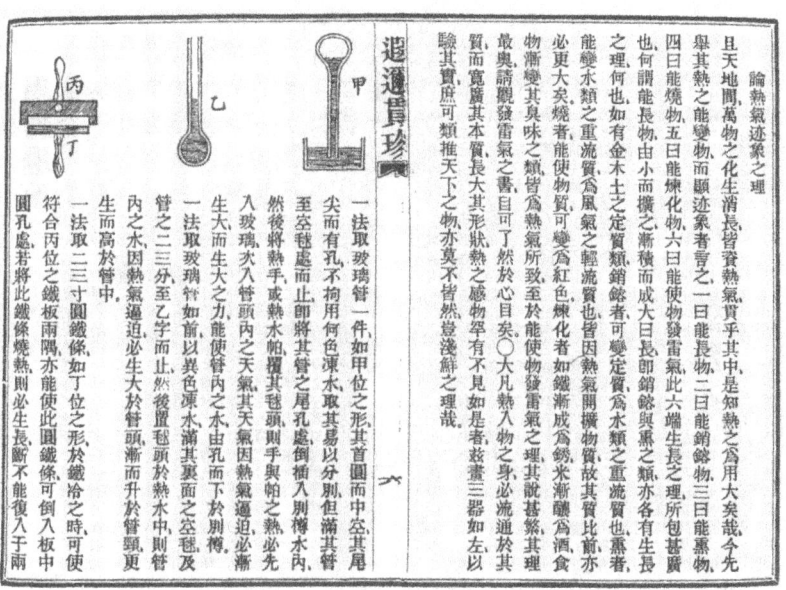

Fig. 3.2 A page from the *Xia'er guanzhen*. Source: Sheng Guowei, Uchida Keiichi, Matsuura Akira (eds.), *Xia'er guanzhen, fu jieti, suoyin* (Shanghai: Cishu chubanshe, 2005), p. 496(223)

Air (*qing liuzhi*, literally "light fluid") is generally named *qi* in Chinese books. The meaning of the word *qi* is the [air] human inhale. Macroscopically, air surrounds the earth. To observe it microscopically, [air] permeates everywhere. Such a thing has forms and properties, and its weight can be measured. Because it flows, hence it is called "light fluid." People do not understand that air is light fluid, and falsely make up terms like metal air, wood air, light air, thunder air, sound air, or even hot air, scent air, or "damp air." [People] just borrow the character *qi* to create those terms. How can air exist in those substances at all? Air is not some groundless, absurd idea but is proven with evidence. It should be classified among [substances such as] metal, wood, water, and earth, but fire is not one of them. Why? Because fire cannot be measured by weight or form. Nor can it be examined according to its intrinsic properties. Without those three characteristics, it cannot be a substance. Fire is a condition of heat, just like wind is the movement of air and wave is the movement of water. … One might ask if air is a substance, in which category of the five phases (*wu xing*) does it belong? The answer is: air has mass and is in effect a substance. It cannot at the same time be metal, wood, water, or earth. As to fire, it is in reality the red flames created by hot metal, wood, water, and earth. … It is just named fire by people. How can it be a substance? Therefore, it is obvious that *wu xing* can be joined by *qi* but fire cannot be listed among them.[34]

In other words, *qi* is the air that humans breathe, and it surrounds the earth. Metal, wood, light, thunder, and heat cannot be air, nor is fire/combustion a substance because it is the consequence, not the cause, of high heat. This explanation redefined the meaning of the words "*qi*" and "fire." However, the term "*wu xing*" was still used as the classification criteria for myriad things.[35]

Therefore, by excluding the metaphysical meaning from the words "*qi*" and "fire," the essays redefined the conceptual framework of understanding thermal phenomena, reducing the influence of traditional natural philosophy. Nevertheless, a few old concepts remained: both sets of texts still used the term "*wu xing*" to denote the concept of myriad things.

In short, the *Bowu xinbian* and the *Xia'er guanzhen* offered their readers rudimentary knowledge about the material understanding of heat, the thermometer, the effects of heat on volume, and heat diffusion. The texts also made great efforts to persuade their readers by redefining "*re*," or "*reqi*," from "heat" and "fire" to "combustion," excluding the traditional Chinese metaphysical meanings from their arguments. In this way, some of the terminological building blocks for explaining the thermal phenomena were shaped, although further refinement was needed and new terms still had to be invented when more advanced texts on heat were written in the future. As mentioned in Chap. 2, the *Bowu xinbian* and probably the *Xia'er guanzhen* had an impact on Xu Shou and Hua Hengfang's work on the steam engine. Although Xu and Hua did not write any technological or scientific treatises, the construction of the *Huanghu* in 1866 was the most eloquent evidence of their transformation from the conventional *gewu qiongli* principle to the modern techno-scientific approach experimentation. Nevertheless, Xu and Hua as well as their fellow Chinese literati faced a more formidable theoretical challenge, which was the birth of thermodynamics.

More Theoretical Challenge: Heat as Motion

While Hobson and the *Xia'er guanzhen* author were negotiating their linguistic and conceptual differences with their Chinese colleagues, the science of heat had moved on. A dynamical theory of heat was challenging the caloric theory. The debate between the material and mechanical theories went back to seventeenth-century natural philosophers' arguments about what exactly the thermometer was measuring. Furthermore, the caloric theory cannot explain why friction creates inexhaustible heat if a given object has a certain heat capacity.[36] Whether heat is a kind of motion or a substance, engineers could still employ the equations that were developed over a century such as calculating specific and latent heat as well as the efficiency of the engine.[37] Furthermore, because heat and light behave similarly—they both radiate over a given distance and transmit through a vacuum—early-nineteenth-century physicists postulated that light, electricity, and heat were vibrations of particles that transmit in undulations or waves, requiring luminiferous aether as the medium.[38]

The wave theory of heat did imply that heat is the vibration of particles, but as a whole, it was soon replaced by the dynamical theory. In the mid-1840s, the British physicist James Prescott Joule (1818–1889), who had deep doubts about the material theory, conducted a series of experiments measuring the ratio between the amount of mechanical work that was destroyed and the amount of heat that was produced through the famous "paddle-wheel" experiment, in which he accurately measured the rising temperature of a pound of water by one degree Fahrenheit. Consequently, he argued that "an amount of *vis viva* (living force) is communicated to it equal to that acquired by a weight of **890** pounds after falling from the altitude of one foot."[39] The experiments, which later led to the formulation of the law of conservation of energy, had an immense impact on physicists, especially William Thomson (1824–1907), professor of natural philosophy at Glasgow University. Thomson had been influenced by Carnot-Clapeyron's theory and, along with his engineer brother James, had been pursuing building an efficient marine steam engine. Joule's discovery created a puzzle for Thomson by revealing that heat is not conserved but destroyed in the process of conversion, which disproved Carnot's perfect cycle theory. The German physicist Rudolf Clausius (1822–1888) offered a solution by arguing that a portion of heat as *vis viva* is converted into work in the process of transferring from high to low temperatures, while the remaining heat flows to the heatsink. Additionally, Thomson's colleague at Glasgow University,

William Rankine (1820–1872), approached the problem similarly to the wave theory of light and argued that, because an atom of matter consists of a nucleus surrounded by an elastic atmosphere that was self-repulsive but attracted by the nucleus, heat, as *vis viva*, constituted the vortex of the atmosphere.[40]

Furthermore, from the 1820s, physicists started to use the ancient term "energy" to denote the ability or property to do work. In the 1850s, William Thomson and his fellow physicists declared that it is energy, not force, that is converted into mechanical work. From 1854, William Thomson (who became Lord Kelvin in 1892) started to use the term "thermo-dynamics," which he defined as "the subject of the relation of heat to forces acting between contiguous parts of bodies, and the relation of heat to electrical agency."[41] Along with other physicists, Thomson, Rankine, and Clausius formulated the concept of energy to replace the conventional term "force" in mechanics.

One should not expect rudimentary knowledge in the *Bowu xinbian* and the *Xia'er guanzhen* to be anywhere near the theoretical sophistication and mathematical complexity offered by Carnot or Thomson. Although there was a certain level of hybridization between Chinese and Western natural philosophies, Western thermal science unquestionably remained dominant because the argument was constructed completely differently from how one might have explained combustion with *qi*, *yin-yang*, and the five phases. It was a step toward a modern understanding of heat. Because of the *Bowu xinbian* and the *Xia'er guanzhen* texts, Chinese readers had the opportunity to learn that cold is the lack of heat and that the level of heat can be measured by a thermometer. Such simple ideas would have played a role in Xu Shou and Hua Hengfang's work on the *Huangbu*.

It should be noted that thermal science was, and still is, an experimental science in which debates on theoretical issues heavily relied on rigorous experimentation. From Antoine Lavoisier, Joseph Black, and James Watt to William Rankine and the Thomson brothers, no one studied the science of heat from an armchair. Even Carnot's highly abstract study on the heat engine was built on the foundation of earlier physicists' experimental works. They carefully designed their experiments, made instruments, measured readings, and confirmed or rejected theories or reformulated new theories. Such a theoretical transformation had a profound impact on the development of the steam engine. Indeed, it changed how engineers approached their engines in almost every aspect.[42]

The discussion above intends to show the complexity of the theoretical development leading to the invention of thermodynamics. Experiments were done to confirm, refute, or reformulate hypotheses that could not have been achieved purely by a textual culture. Furthermore, their works were done in institutional settings where universities, learned societies of scientists or engineers, journals, and regular meetings were essential opportunities for knowledge to be passed on from teachers to students and new ideas to be exchanged between pioneering physicists. Such institutional settings had not appeared in China in the 1850s.

In the 1860s, the Qing government started to set up new schools to teach the subjects of science and mathematics. The School of Combined Learning (*Tongwen guan*), a school originally established by the *Zongli yamen* for training translators and interpreters in Beijing in 1862, was the first to offer such courses, and it used a set of textbooks written by one of its teachers, W.A.P. Martin.

TEACHING HEAT: THE *GEWU RUMEN*

W.A.P. Martin, who had a university degree in natural philosophy at Indiana University in the United States, started to consider introducing modern science to non-Western people as an effective means of proselytization when he was training in the seminary in the 1830s.[43] He gained a good command of Chinese while doing missionary works in Ningbo, Zhejiang province, from 1850. Subsequently, he composed religious tracts and translated the Bible into Chinese.[44] During the Second Opium War, he was twice recruited by the U.S. envoy to China as an interpreter in treaty negotiations. It was then that he had contacts with Qing government officials and was invited by Qing officials to translate Henry Wheaton's *Elements of International Law*, a university-level textbook in international law first published in 1836. The end result was the *Wan'guo gongfa*, which was published under the *Zongli yamen*'s sponsorship in 1864.[45]

When he was working on the *Wan'guo gongfa*, Martin was considering writing a natural philosophy textbook in Chinese. He felt that the Chinese education system created scholars who only focused on moral philosophy and politics but knew as little about nature as the Europeans before the Renaissance. He especially criticized the conception of *yin-yang* and the five phases:

> With [the Chinese] levity is a force as real as gravity; cold and darkness no less than light and heat. They find a ready explanation for all phenomena in the 'play of dual forces'; *Yin yang kiao kan* [*yinyang jiaogan*, the interactions between *yin* and *yang*] is a formula as good to hide ignorance as many a phrase in vogue with us. Their chemistry has not emerged from the chrysalis of alchemy. They count five elements instead of our ancient four—metal being added, and wood taking the place of air, which is omitted as too subtle to suit their idea of substance.

Therefore, by 1860, Martin had planned to write a book about natural philosophy to expose the fallacy of the traditional Chinese conceptions of "dual forces" (*yin* and *yang*), "five elements" (five phases), and "*fengshui*" (geomancy).[46]

In 1865, Martin was recruited by the *Zongli yamen* into Beijing's School of Combined Learning as Professor of English but later in the same year was moved to the position of Professor of International Law.[47] In 1867, to improve his knowledge on the subject, he went back to the United States to study at Yale University for two years. While at Yale, he spent his spare time writing a textbook about natural philosophy composed in the format of catechism. In 1869, he was appointed principal of the School of the Combined Learning. At roughly the same time, he had reached the final stage of completing his manuscript. He then gained *Zongli yamen* officials' assistance in polishing his Chinese prose. In 1868, he published the book as the *Gewu rumen* (*An Introduction to Natural Philosophy*), which became the textbook for the School's science courses.

The *Gewu rumen* is comprised of seven volumes that cover the subjects of pneumatics (*Qixue*, the Study of Air), hydraulics (*Shuixue*, the Study of Water), heat and optics (*Huoxue*, the Study of Fire), electricity (*Dianxue*, the Study of Electricity), mechanics (*Lixue*, the Study of Force), chemistry (*Huaxue*, the Study of Change), and computation (*Suanxue*, the Study of Computation), which is the mathematical exposition of the previous six volumes, not the subject of arithmetic. These are common subjects in nineteenth-century natural philosophy textbooks published in Britain and the United States.[48] According to its preface, Martin wrote the book on the basis of various sources.[49] One of them could have been *Handbook of Natural Philosophy* by Dionysius Lardner, natural philosophy professor at University College London between 1828 and 1831. *Handbook of Natural*

Philosophy was first published in London in 1852 and became a popular textbook in Britain and the U.S. It has similar subjects to the *Gewu rumen*. If Martin did consult Lardner's *Handbook of Natural Philosophy*, he selected and simplified the contents to fit the needs of the Chinese students at the School of the Combined Learning. We will use the 1853 Philadelphia edition of the *Handbook of Natural Philosophy*, which was the reprint of the 1852 London edition, for comparison.

Before we turn our attention to the *Gewu rumen*'s volume on heat, we need to discuss the second volume, the *Qixue*. The *Qixue* has three chapters: "On Air (*Lun tianqi*)," "On Steam (*Lun zhengqi*)," and "On Sound (*Lun yinsheng*)." It denotes air or atmosphere with the same term, *tianqi* (literally "air of the sky"), without making a distinction between the two. In essence, the "On Air" chapter introduces the concept of atmosphere and atmospheric pressure (*tianqi xiaya*). It also presents the barometer, which measures atmospheric pressure and hence can predict weath, and the cylinder-piston mechanism of the air pump, which can pump air out of a vessel. The contents of this chapter, though similar, are better structured and explained by the *Bowu xinbian*.

Does Heat Contain Force?

The "On Steam" chapter outlines four principles to explain why steam could be turned into mechanical work: first, heat contains more "force" (*li*) than cold; second, the hotter the steam, the greater the force; third, the higher the steam pressure, the greater the force; fourth, water has to be pressurized to reach an extremely high temperature. In that context, the word "*li*" reads like energy. However, the selection of the word "*li*" to denote energy is quite confusing. *Li* is an ancient idea that could mean muscle strength, power, or ability in Chinese, but at that time, the concept of "energy" had not been introduced into the Chinese language.

In the *Gewu rumen*'s *Lixue* (mechanics) volume, *li* (force) is defined as the cause of a body's motion. A force can cause a body in motion to stop or a motionless body to move, and the measurement of "*wu dong zhi li*" ("the force that moves a body") is weight (mass) times speed; this is the concept of momentum.[50]

The confusion between "force" and "energy" might not have been peculiar among non-specialists in the nineteenth century because the word "energy" was already being used in everyday nontechnical language. For example, the "energy" entry in an English dictionary published in the

1880s reads, "Energy, ... power; force; the power of operating or doing; vigorous action; efficacy; spirit; life."[51] Therefore, Martin's problem was finding a Chinese term for energy. Before such a term was invented and popular science texts or primary education textbooks were written to educate the general public about energy, those who were not trained as physicists or engineers would likely associate energy with power or force.

Regardless of the confusion between force and energy, the chapter explains that it is steam's expansive properties that exert a force on the piston and hence drive the steam engine. Superheated high pressure steam is more powerful than steam under atmospheric pressure. The *Gewu rumen* also gives a simple history of the steam engine from Newcomen to Watt, then provides a detailed description of the engine and how the Watt engine was used as the source of power to drive cotton spinning and weaving machinery as well as steam locomotives and steamboats. It also provides illustrations of these machines.[52] In other words, the "On Steam" chapter gives a comprehensive explanation of the principle of the steam engine through the idea *li*. This was a much better explanation than either the *Xia'er guanzhen* or the *Bowu xinbian* offered.

Now, we will return our attention to the *Huoxue*. This volume has two chapters: *Reqi* (Heat, literally "hot air") and *Guangxue* (Optics).

The *Reqi* chapter has eighty-two sets of questions and answers, which can be roughly divided into six sets of concepts: the nature of heat, the thermometer, heat diffusion, specific heat, the effect of heat on state, and the relationship between heat and force. Thirteen illustrations of instruments, including different types of thermometer and calorimeter, were included after the table of contents.

The opening set of questions and answers explains what heat is and its distinction from fire:

> [Heat] is invisible. It is silent, and its body is minuscule and difficult to test... Heat permeates in myriad things; without it, wind does not blow, water does not flow, humans and animals do not live, nor do grasses and woods grow. Heaven and earth would be dead bodies [without it].

Describing that heat is invisible and silent is a literary expression of stating that heat is imponderable. Then, the text states that the sources of heat can be electricity, friction, and chemical reactions. It stresses that heat and light are not the same thing:

3 TRANSLATING HEAT: TACKLING OLD CONCEPTIONS WITH NEW IDEAS... 77

[Heat and light] may coexist but are not necessarily dependent on each other. There can be heat without light, or light without heat. Even if they coexist, one can separate them. For example, one can use glass to separate heat from firelight because light passes through glass but heat cannot. [One can] use an iron sheet to separate light from heat because heat passes through the iron sheet but light cannot. However, sunlight is slightly different from firelight because both heat and light [from the sun] can pass through glass.

The following question asks what fire is:

[Fire] is the manifestation of heat. An extremely hot body emits light. The [chemical] combination of two things can start a fire. Just as the reason why wood combusts is that life-supporting air [oxygen] in the wind bonds with carbon essence [carbon] in the wood, and hence a fire starts. Irons or rocks could glow when they are heated red hot by fire, but they in essence cannot combust.[53]

The meaning of the first sentence is unclear because an extremely hot body does not necessarily combust, although it may emit light, a phenomenon now known as incandescence. Unfortunately, the text provides no explanation for why this is so.

Next, it explains the difference between heat and cold:

[If one touches a] body [that is colder than one's hand], heat would enter the body from one's hand. Because the heat in one's hand decreases, therefore [one] feels cold. [If one touches a] body [that is hotter than one's] hand, then heat would enter one's hand [from the body]. Because heat in the hand increases, hence [one] feels hot. To infer from such phenomena, [we know that] there is just heat [but no cold]. The absence of heat is cold, and there is no such thing as "cold." [It is] similar to light, which is a thing. The absence of light is darkness. There is no such thing as darkness other than light.[54]

In other words, hot or cold sensations are the result of the flow of heat in or out of the human hand. The analogy of light was more logical than the *Bowu xinbian*'s analogy of flavor, which is itself a sensation.[55]

To discuss the effect of heat on volume, the text gives descriptions of a few simple experiments, such as putting a glass bottle of cold water in hot water and observing how the level of the cold water rises in the bottle or observing the changes in dimension of two fitted iron blocks.[56] Then, it describes the numeral differences in the thermal expansion and contraction of various materials; for example, silver would expand from a length of 3.2 *chi* (roughly 3.5 feet) to 5.2 *chi* (roughly 5.7 feet), but gold to 6.8 *chi* (roughly 7.4 feet).[57] These descriptions fit the purpose of education well, although one might argue that they do not explain why heat can create changes in volume.

Next, the text introduces the thermometer, which it names the *hanshu biao* (literally "cold and heat scale"). This term is similar to the *Bowu xinbian*'s *hanshu zhen*. Both terms convey the idea that a thermometer is an instrument for measuring the level of hot and cold. The text gives a simple description of a mercury thermometer and highlights the fixed point of 32 degrees Fahrenheit as the freezing point of water.[58] Then, the text introduces heat diffusion (*reqi fasan*) from the calorist point of view, explaining that everything contains different levels of heat and heat is like water that flows from higher to lower levels. For example, if a bar of hot iron was placed on top of a bar of cool one, the hot one would become cooler but the cool one would become hotter until they reached the same level of heat. Hence, the heavens and the earth have a constant amount of heat from the ancient times to today. (This idea was known to be the conservation of heat.) Without further explaining this concept, the text then gives a very short description of heat conduction and radiation, and a short explanation that the speed of heat conduction might be different.

Introducing Specific Heat

The *Reqi* chapter also discusses specific heat, a concept that is not mentioned in either the *Bowu xinbian* or the *Xia'er guanzhen*. It uses the term *reliang* (literally "the amount of heat") to denote the idea. The text states that different substances have different capacities to absorb heat. If one hammers a bar of cool iron and reduces its size, the iron would become hotter because the reduction of its volume would force the heat it contains to spill over. This idea was a calorist view that the reduction of the volume of a substance decreases its capacity to absorb heat and vice versa.[59] Hence, Martin's explanation assumes that heat, as a fluid contained within the iron bar, can be forced out by percussion.[60] This idea became obsolete after the dynamical theory of heat was invented in the mid-nineteenth century, which argues that the hammering converts kinetic energy to thermal energy and hence increases the temperature of the body.

Next, the section gives descriptions of experiments designed to increase temperature. It explains that a tael of water requires about 30 times as much heat as a tael of mercury to raise its temperature by one degree Fahrenheit. Specifically, it describes the method of measuring specific heat by immersing two vessels, one containing a tael of mercury and the other the same weight of water, into boiling water until both are heated. Then they are put into an ice hole to measure the weight of the condensed water on the surface of the vessels. The water on the mercury vessel is thirty times the weight of water on the surface of the water vessel.[61]

Subsequently, the *Reqi* chapter explains that the length of daytime may cause seasonal changes and discusses how heat changes the state of matter.[62] It argues that an increase of heat reduces the attraction between the particles of matter and hence causes liquidation, while the reduction of heat solidifies matter.[63] The text also explains that melting solids requires more heat and gives a description of three experiments. In the first experiment, salt and ice are combined, and the ice melts into water, but the mixture is actually colder than ice itself. Unfortunately, the chapter does not explain why this occurs.[64]

Refusing to Accept the Dynamic Theory

The last section of the *Reqi* chapter introduces the concept of convertibility between "*li*" and heat (*li re husheng*, the mutual production of force and heat), by which it means the convertibility between mechanical work and heat.[65] The chapter explains that "scientists have argued that heat is not a substance but the consequence of the vibration of particles. Heat, which increases the warmth [the temperature] of one pound of water by one degree [Fahrenheit], could raise a pound of stone 772 feet high. Alternatively, the force generated by the dropping of one pound would generate heat, increasing a pound of water by one degree [Fahrenheit]."[66]

However, Martin was not sure whether or not the mechanical equivalence of heat was an accurate theory. In the final section of the *Reqi* chapter, he compares the material and mechanical theories of heat:

> It is difficult to tell. Both theories have proofs, but neither of them fully explains every phenomenon. If heat is an imponderable substance, it permeates every object, and the increase of heat is similar to the rise of water. If one uses force to make a body suddenly contract, the body will become hotter. It is like water being pressed out of a body. If [one] argues that heat

> is not a substance, how [does one explain] such a phenomenon? Contending that heat is not a substance is no different from contending that light is not a substance. The vibration of air hits ear and hence [one can hear] sound. It is the vibration of the air. It is not like shooting an arrow, which [flies] from far-away to near-by. The transmission of sound needs air, and the transmission of heat requires a different kind of mysterious air (*xuanqi*), which permeates heaven and earth. Light, heat, and electricity all require this ether. The theory seems to be right, but if one compares [the two theories] on the basis of [the phenomenon] of hot air, the material theory seems to be a more convenient one.[67]

That is to say, in Martin's opinion, the material theory of heat was more reasonable than the dynamical theory in explaining thermal expansion and contraction as well as why percussion increases a body's temperature. However, this passage also shows the influence of the wave theory, which in itself suggested the vibration of particles. It seems that Martin did not give much thought to that suggestion.[68]

In translating luminiferous aether, Martin's selection of the term "*xuanqi*", which could mean "mysterious air" or, according to the *Book of Han* (*The History of the Former Han*), "the heavenly *qi*", shows the influence of his Chinese collaborator, who would have known the *qi* theory quite well. In other words, Martin's understanding of heat was predominately material but was colored by the wave theory.

We do not know why Martin still held on to the old caloric theory. One of Martin's source texts, Dionysius Larnder's *Handbook of Natural Philosophy*, states:

> Closely connected with the subject of the quantitative measurement of heat, there is a principle of great practical importance, which has long been conjectured to be true, but which has only within the last few years been proved experimentally, that *all the different forms of physical energy, whether chemical action, light, heat, electricity, magnetism, or visible motion and mechanical power, are convertible into each other* [original emphasis]; and that, although physical energy may be converted from one form to another, or transferred from one portion of matter to another, its whole amount in the universe in unchangeable.[69]

Therefore, although Martin had a university degree in natural philosophy, he could not keep up with the development of physics after he went to China. Although he might be able to write a textbook by drawing on materials from the latest sources, he probably still firmly believed in what he had learned at university in the 1830s.

In short, W.A.P. Martin's *Gewu rumen* aimed to impart basic knowledge of nineteenth-century physics and chemistry to Chinese students who were under the influence of traditional *qi* and five-phase theory. In plain language, it provided a clear explanation of issues including the nature of heat, the thermometer, and the effect of heat on volume. It also taught basic ideas of heat capacity. However, it did not fully embrace the mechanical theory of heat as a credible explanation for the nature of heat.

It should be noted that experimentation is an essential part of the *Gewu rumen*. In explaining every concept, it tells readers how to conduct an experiment to prove the point. Experimentation, in which hypotheses are proved or disproved by controlling variables and taking exact measurements, was not considered an essential part of observing nature at the time. The advance of thermal science from seventeenth-century alchemists' phlogiston and Antoine Lavoisier's caloric to Joule's mechanical equivalence of heat would not have happened without devising controlled experiments to test hypotheses. Unfortunately, except for the fields of medicine and alchemy, both of which relied on the conceptions of *qi*, *yin-yang*, and the five phases, China lacked a strong tradition of scientific experimentation.

In the School of Combined Learning, teachers would have used Martin's textbook in their science classes and guided their students to conduct experiments. This method of teaching was new to Chinese students, who were familiar with memorizing texts and writing essays. The number of science students at the School of Combined Learning was small. In 1879, there were only 66 total in the classes of astronomy, chemistry, mathematical physics, and physiology. Nevertheless, the school marked a new beginning in science education in China.[70]

Conclusion

In the mid-1850s, Western missionaries in China published the popular science book *Bowu xinbian* and essays in the magazine *Xia'er guanzhen* as part of their efforts to convert the Chinese people to Christianity. These texts provided rudimentary knowledge about the material theory of heat, temperature, and the effects of heat on volume. This simple knowledge was the first opportunity for the Chinese literati to learn about an explanation of thermal phenomena that offered an alternative to traditional conceptions of *qi* and the five phases. W.A.P. Martin's *Gewu rumen*, which was published in 1868 as a textbook at the School of Combined Learning in Beijing, was better structured than these earlier works and provided clearer elucidations of heat. Similar to Hobson and the author of the *Xia'er guanzhen*, Martin made an effort to distinguish *qi* and heat. He also introduced the terms "*hanshu biao*" to denote the thermometer and "*reliang*" to denote specific heat. Nevertheless, Martin used the term "*li*" ("force") to translate "energy" and hence confused energy with force. Because Martin's *Gewu rumen* was used as a textbook in the School of Combined Learning, both its detailed explanations and the descriptions of experiments would have brought the School of Combined Learning students a better understanding of thermal phenomena.

Neither Hobson nor Martin could produce anything close to the theoretical depth and mathematical technicality of Carnot or Clausius, but the contents of their works were a step toward a new understanding of thermal phenomena, which were fundamental in modern science and technology. Considering that by the nineteenth century, the Chinese literati could only explain combustion and heat through *qi* and its five phases, these texts provided Chinese readers, regardless of their backgrounds, with a glimpse of how to understand thermal phenomena from a new perspective. Instead of rising up against the publication of such alien knowledge, the Chinese literati devoured them. It is not clear how much impact the *Xia'er guanzhen* essays made, but the *Bowu xinbian* was part of Xu Shou and Hua Hengfang's efforts to experiment on the steam engine, and it remained a popular Western learning text by the 1890s.

Similarly, the *Gewu rumen*'s influence went beyond the School of Combined Learning. It was used in many missionary schools.[71] The Chinese literati and Qing government officials had read it, and its contents were reprinted in the missionary newspaper *Jiaohui xinbao* (*Church News*) in Shanghai.[72] It was even reprinted in Japan and became one of the physics textbooks at Japan's naval school.[73] Indeed, it was so popular that Martin did not think of revising it until 1888, when, under the recommendation of government officials, he made some revisions to the textbook's language and illustrations without changing much of its content. He ultimately presented the new edition to Emperor Guangxu (r. 1875–1908).[74] In other words, by the 1890s, the *Gewu rumen* had become one of the authorities of Western scientific knowledge in China.

It is worth noting that Martin finally accepted the dynamical theory of heat. At the beginning of the 1888 edition of the *Huoxue*, Martin states that "What is *reqi* (heat)? It is not *qi*, but actually the transformation from *li* (force)."[75] However, in the fast-progressing field of thermal physics in the second half of the nineteenth century, Martin's change of heart was too late. The *Gewu rumen*'s contents had already been outdated even by the 1860s standard.

The next chapter will discuss other texts about heat that were translated into Chinese and how the Chinese literati might have understood them.

Notes

1. I find that over-emphasizing the issue of incommensurability, which is championed by Thomas Kuhn and Paul Feyerabend, in studying science translation limits our understanding of science in translation. G.E.R. Lloyd reminds us that there must be some common understanding between different epistemological systems, however dissimilar they are. Thomas S. Kuhn, *The Structure of Scientific Revolution* (Chicago: Chicago University Press, 1962); Paul K. Feyerabend, "Explanation, Reduction and Empiricism," in H. Feigl and G. Maxwell (ed.), *Scientific Explanation, Space, and Time*, Minnesota Studies in the Philosophy of Science, vol. 3 (Minneapolis: University of Minneapolis Press, 1962), pp. 28–97; G.E.R. Lloyd, *Ancient Worlds, Modern Reflections: Philosophical Perspectives on Greek and Chinese Science and Culture* (New York: Oxford University Press, 2004). Compare Lydia H. Liu (ed.), *Tokens of Exchange: The Problem of Translation in Global Circulations* (Durham, Duke University Press, 1999).
2. The Song dynasty Confucian scholar Zhu Xi (1130–1200) advocated the *gewu qiongli* principle in self-cultivation and understanding the world. The principle became the foundation of the Chinese world view for the next six centuries. Yung Sik Kim, *The Natural Philosophy of Chu Hsi (1130–1200)* (Philadelphia: American Philosophical Society, 2000).
3. The mutual-production-and-mutual-conquest relationship is: wood produces fire, fire produces earth, earth produces metal, metal produces water, and water produces wood. However, wood conquers earth, metal conquers wood, fire conquers metal, water conquers fire, and earth conquers water. This idea was raised in the second century B.C. book *Huainanzi* (*The Masters of Huainan*), a collection of essays that resulted from scholarly discussions and debates in the court of Liu An, the Prince of Huainan of the Han Dynasty. Liu An, *Huainan honglie jijie*, edited by Liu Wendian, (Beijing: Zhonghua, 1989), vol. 1, p. 124, 146.

4. I am not an expert in traditional Chinese cosmology and natural philosophy, but I have learned a lot from Nathan Sivin, *Traditional Chinese Medicine in Contemporary China* (Ann Arbor: University of Michigan Centre for Chinese Studies, 1987); Joseph Needham, *Science and Civilisation in China*, Volume 4: Physics and Physical Technology, Part 1: Physics (Cambridge: Cambridge University Press, 1962); John B. Henderson, *The Development and Decline of Chinese Cosmology* (New York: Columbia University Press, 1984); Christopher Cullen, "The Science/Technology Interface in Seventeenth-Century China: Song Yingxing on '*qi*' and the '*wu xing*'," *Bulletin of the School of Oriental and African Studies, University of London* 53:2 (1990), pp. 295–318; Xi Zezong (ed.), *Zhongguo kexue jishu shi: kexue sixiang juan* (Beijing: Kexue chubanshe, 2001); Dai Nianzu (ed.), *Zhongguo kexue jishu shi: Wulixue juan* (Beijing: Kexue chubanshe, 2001); Onozawa Seiichi, Fukunaga Mitsuji, Yamanoi Yū (eds.), *Qi de xixiang: Zhongguo ziranguan yu ren de guannian de fazhan*, translated by Li Qing (Shanghai: Shanghai renmin chubanshe, 1990).

5. One could dip into the "Fire" chapter of the famous sixteenth-century *Bencao gangmu* (*Compendium of Materia Medica*) by the Ming dynasty herbalist and physician Li Shizhen to see how one might employ the *qi*, *yin-yang*, and the five phases to explain heat and combustion through classification. In essence, the category of *yang* or fire, which is strong *yang*, decides whether a given material is hot or combustible. A good discussion can be found in Li Jianmin, "*Bencao gangmu* Huo bu kaoshi," *Bulletin of the Institute of History and Philology, Academia Sinica* 73:3 (2002), pp. 395–441.

6. Song Yingxing, *Ye yi; Lun qi; Tan tian; Si lian shi* (Shanghai: Shanghai renmin chubanshe, 1976, reprint of seventeenth century editions), p. 82. For a more detailed discussion of Song Yingxing's philosophy, see Christopher Cullen, "The Science/Technology Interface in Seventeenth-Century China: Song Yingxing on '*qi*' and the '*wu xing*'," *Bulletin of the School of Oriental and African Studies, University of London* 53:2 (1990), pp. 295–318.

7. J. R. Partington and D. MacKie, *Historical Studies on the Phlogiston Theory* (New York: Arno Press, 1981).

8. Chen Wancheng (Chan Man Sing), Luo Wanwei (Law Yuen Mei), Kuang Yongheng (Kwong Wing Han), "Wan Qing xiyixue de yishu: yi *Xiyi luelun, Fuying xinshuo* liangge gaoben weili," *Journal of Chinese Studies, The Chinese University of Hong Kong.* 56 (2013), p. 244. The historian Zou Zhenhuan has also examined the background of the *Bowu xinbian*; see Zou Zhenhuan, "He Xin ji qi bianyi de '*Bowu xinbian*'," *Shanghai keji fanyi* 4:1 (1984), p. 45.

9. Robert A. Bayliss and C. William Ellis, "Neil Arnott, F.R.S., Reformer, Innovator and Popularizer of Science, 1788–1874," *Notes and Records of the Royal Society of London* 36:1 (1981), pp. 103–123.
10. Chen Wancheng (Chan Man Sing), Luo Wanwei (Law Yuen Mei), Kuang Yongheng (Kwong Wing Han), "Wan Qing xiyi xue de yishu: yi *Xiyi luelun, Fuying xinshuo* liangge gaoben weili," p. 245.
11. John Fryer, "An Account of the Department for the Translation of Foreign Books" *North China Herald*, 29 Jan 1880, pp. 77–81. The Chinese edition of the same account can be found in Fu Lanya (John Fryer), Jiangnan zhizao zongju fanyi xishu shilüe (A Summary of the translations of the Jiagnan Arsenal), quoted in Xiong Yuezhi, *Xixue dongjian yu wan Qing shehui yu wan Qing shehui*, p. 394.
12. Chan Man Sing, "Sinicizing Western Science: The Case of *Quanti xinlun*," pp. 528–556; Chen Wancheng (Chan Man Sing), Luo Wanwei (Law Yuen Mei), Kuang Yongheng (Kwong Wing Han), "Wan Qing xiyixue de yishu: yi *Xiyi luelun, Fuying xinshuo* liangge gaoben weili," pp. 243–292.
13. Wang Yangzhong, "Hexuli *Kexue daolun* de liangge Zhongyi ben: Jiantan Qingmo kexue yizhu de zhunquexing," *Zhongguo keji shiliao* 21:3 (2000), pp. 207–221; Iwo Amelong, "Some Notes on Translations of the *Physics Primer* and Physical Terminology in Late Imperial China, *Wakumon* 11:8 (2004), pp. 11–33.
14. In 1774, the British chemist Joseph Priestley, a firm phlogiston believer, discovered a new kind of air by using a "burning lens" to focus sunlight on a lump of "mercury calx" (mercuric oxide) and collected a new kind of air, which he called dephlogisticated air. Upon hearing of Priestley's discovery, Lavoisier repeated Priestley's experiment and, from 1775–1780, formulated a new theory by reversing the phlogiston argument. He contended that the chemical combination between the element of oxygen in the air and combustible materials causes combustion. However, he still considered heat a substance, which he called "caloric." Although the majority of nineteenth-century chemists and physicists accepted the idea of caloric, they did not completely agree on whether caloric was composed of atoms. Specifically, John Dalton (1766–1844), who pioneered the modern atomic theory, postulated that an atom has a heavy central particle surrounded by an atmosphere of caloric, and the size of an atom is decided by the diameter of the caloric atmosphere. Joshua C. Gregory, *Combustion from Heracleitos to Lavoisier* (London: Edward Arnold & Co. 1934); James Bryant Conant (ed.), "The Overthrow of the Phlogiston Theory: The Chemical Revolution of 1775–1789," in idem. (ed.), *Harvard Case Histories in Experimental Science*, vol. 1 (Cambridge, Mass.: Harvard University Press, 1966), pp. 65–116; Robert J Morris, "Lavoisier and the Caloric Theory," *The British Journal for the History of Science* 6:1 (1972),

pp. 1–38. For Dalton's theory, see Robert Fox, "Dalton's Caloric Theory," in D.S.L. Cardwell (ed.), *John Dalton and the Progress of Science* (Manchester: Manchester University Press, 1968), pp. 109–124.
15. For more details, see Duane Roller, "The Early Development of the Concepts of Temperature and Heat: The Rise and Decline of the Caloric Theory," in James Bryant Conant (ed.), *Harvard Case Histories in the Experimental Sciences*, vol. 1, (Cambridge, MA: Harvard University Press, 1957), pp. 117–214.
16. The seventeenth-century chemist Robert Boyle (1627–1691) had famously used an air pump to experiment with the heating of a gas by mechanical compression and cooling by expansion. Although late-eighteenth and early-nineteenth-century chemists and physicists did not agree whether the expansion or contraction of a body would increase or decrease its heat capacity, they postulated that the amount of caloric contained in a body is its capacity for heat. Each particular substance takes a certain amount of heat, and that amount depends on the density of the material. A complete lack of caloric would be absolute zero temperature. For more details on the history of the concept of heat capacity, see Hasok Chang, *Inventing Temperature: Measurement and Scientific Progress* (Oxford: Oxford University Press, 2004), pp. 65–66, 168–169.
17. Watt worked at the University of Glasgow as an instrument maker and was asked to repair a model Newcomen engine. Joseph Black was a professor of medicine at the same university. As such, Watt was aware of Black's discovery of latent heat.
18. Watt's cut-off technique involved using the indicator diagram, which plotted the state of steam in the cylinder by pressure and volume. Such a diagram was an effective visual tool for technicians to control the supply of steam to the cylinder. Watt also invented "horsepower," which is the unit of lifting 33,000 lbs. one foot high per minute, to measure the mechanical work of his engine. D.S.L. Cardwell, *From Watt to Clausius: The Rise of Thermodynamics in the Early Industrial Age*, pp. 40–55.
19. For more details, see H.W. Dickinson, "The Steam-Engine to 1830," in Charles Singer, E.J. Holmyard, A.R. Hall and Trevor L. Williams (eds), *A History of Technology*, vol. 4: The Industrial Revolution, c1750 to c1850 (Oxford: Oxford University, 1958), pp. 168–198; Richard L. Hills, *Power from Steam: A History of the Stationary Steam Engine* (Cambridge: Cambridge University Press, 1989), pp. 1–161.
20. Carnot explains that an ideal heat engine runs in cycles, which are composed of four processes: isothermal expansion, adiabatic expansion, isothermal compression, and adiabatic compression. See D.S.L. Cardwell, *From Watt to Clausius*, p. 52; Sadi Carnot, *Reflections on the Motive Power*

of Heat, edited and translated by Robert Henry Thurston (New York: J. Wiley & Sons, 1890).
21. Clapeyron used the pressure-volume diagram, which was derived from Watt's indicator diagram. In his mathematical formulation of the Carnot cycle, the net mechanical work is the area enclosed by the cycle described by Carnot. Because the lines of the cycle are curved, infinitesimal calculus is necessary in the calculation.
22. For a succinct discussion of the birth of classical thermodynamics, see Hasok Chang, "Thermal Physics and Thermodynamics," in Jed Buchwald and Robert Fox (eds), *The Oxford Handbook of The History of Physics* (Oxford: Oxford University Press, 2013), pp. 473–507. A more detailed discussion can be found in D.S.L. Cardwell, *From Watt to Clausius*.
23. Neil Arnott, *Elements of Physics*, vol. 2, (London: Longman, Rees, Orme, Brown, and Green, 1833), pp. 73–75.
24. *Hanshu* can also mean the span of time between summer and winter, implying a whole year.
25. He Xin, *Bowu xinbian*, 1:8ab.
26. He Xin, *Bowu xinbian*, 1:21b.
27. He Xin, *Bowu xinbian*, 1:22ab.
28. Neil Arnott, *Elements of Physics*, v. 2, part 1, p. 34.
29. He Xin, *Bowu xinbian*, 1:23ab.
30. For more discussion of the *Bowu xinbian*'s intrinsic heat, see Feng Shanshan, "Jindai xifang rexue zai Zhongguo de chuanbo (1855–1902)," PhD thesis, Inner Mongolia University, (2019), p. 118.
31. Shen Guowei et al. (eds.), *Xia'er guanzhen, fu jieti, suoyin* (Shanghai: Cishu chubanshe, 2005), p. 497(222).
32. There was no equivalent term to denote "the state of matter," hence, it used the phrase "*wu zhi xingshi*," which literally means "the forms and conditions of things."
33. Shen Guowei et al. (eds.), *Xia'er guanzhen, fu jieti, suoyin*, pp. 497(222)–496(223).
34. Shen Guowei et al. (eds.), *Xia'er guanzhen fu jieti suoyin*, p. 468 (251).
35. Such usage was common in Chinese classics. For example, in the *Book of Documents* (*Shang shu*), the term "*wu xing*" is used to indicate "everything people need in their daily lives."
36. The dynamical theory of heat can be traced back to the early seventeenth century when Galileo Galilei (1564–1642) was working on a thermometer that measured the temperature of the human body. He argued that the human sensation of heat is caused by rapid motions of certain atoms and hence is just an illusion of the human senses. René Descartes (1596–1650), who thought of all physical phenomena in terms of extensions and motions, believed that the world is composed of three kinds of particles and that the

vibration of the infinitely small fire particle is responsible for heat as well as thermal expansion and contraction. Robert Boyle argued that heat is the motion of constituent atoms. Although the material theory of heat was dominant after the late eighteenth century, in 1798, Benjamin Thompson (1753–1814), an American who fought on the British side during the American Revolutionary War and later moved to Bavaria to serve as an advisor to the Bavarian king, found that boring canons create enormous heat that can boil water. As such, he argued that the source of heat is motion, not caloric. D.S.L. Cardwell, *From Watt to Clausius*, pp. 2–5, 97–107.

37. E. Mendosa, "A Sketch for a History of Early Thermodynamics," in Spencer Weart and Melba Phillips (eds), *History of Physics* (New York: American Institute of Physics, 1985), pp. 18–24.
38. Stephen G. Brush, 'The Wave Theory of Heat: A Forgotten Stage in the Transition from the Caloric Theory to Thermodynamics', *The British Journal for the History of Science* 5:18 (1970), pp. 145–167.
39. Joule conducted several experiments to prove his conviction that heat and mechanical work is convertible. The most well-known of these experiments involved a paddle-wheel placed in a can that was filled with water. Weights were attached over pulleys working in opposite directions, turning the paddle-wheel. The idea of *vis viva* or living force was invented by the seventeenth-century philosopher and mathematician Gottfried Leibniz (1646–1716), who argued that the "force" of a moving body, which he called "*vis viva*," can be transferred from one body to another. He defined the value of *vis viva* as mass times velocity squared. For more details, see Cardwell, *From Watt to Clausius*, pp. 231–238; Crosbie Smith, *The Science of Energy: A Cultural History of Energy in Victorian Britain*, pp. 71–72.
40. D.S.L. Cardwell, *From Watt to Clausius*, pp. 239–294; Crosbie Smith, *Science of Energy*, pp. 86–95.
41. William Thomson, "On the Dynamical Theory of Heat," *Transactions of the Royal Society of Edinburgh* 21, part 1, (1853), p. 123.
42. Lynwood Bryant, "The Role of Thermodynamics in the Evolution of Heat Engines," *Technology and Culture* 14:2 Part 1 (1973), pp. 152–165.
43. Peter Duus, "Science and Salvation in China: the Life and Work of W.A.P. Martin (1827–1916)," in Kwang-ching Liu (ed.), *American Missionaries in China: Papers from Harvard Seminars* (Cambridge, MA: Harvard East Asian Research Center, 1970), pp. 11–41; Ralph R. Covel, *W.A.P. Martin, Pioneer of Progress in China* (Washington DC: Christian University Press, 1978); Fu Deyuan, *Ding Weiliang yu jindai Zhong Xi wenhua jiaoliu* (Taipei: Taiwan University Press, 12013).
44. Martin participated in a project of translating the Bible into the Ningbo dialect and teaching in a missionary school. His excellent ability in Chinese

composition was shown by the *Tiandao suyuan* (Tracing Back to the Origin of the Heavenly Way), which he published in 1854. It became one of the most popular missionary tracts in nineteenth-century China. Soon, he started to write *The Analytical Reader: A Short Method for Learning to Read and Write Chinese* to teach Chinese to foreign missionaries. It was published in Shanghai by the Presbyterian Mission Press in 1863.

45. For further discussions, see Lydia H. Liu, "Legislating the Universal: The Circulation of International Law in the Nineteenth Century," Lydia H. Liu (ed.), *Tokens of Exchange: The Problem of Translation in Global Circulations*, pp. 127–164; Lydia H. Liu, "Translating International Law," in idem., *Clash of Empires: The Invention of China in Modern World Making* (Cambridge, Mass.: Harvard University Press, 2006), pp. 108–139.
46. W.A.P. Martin, *A Cycle of Cathay or, China, South and North with Personal Reminiscences* (New York: Fleming H. Revell Company, 1900), p. 236.
47. For more details on the *Tongwen guan*, see Knight Biggerstaff, *The Earliest Modern Government Schools in China* (Ithaca, N.Y.: Cornell University Press, 1961).
48. In the *Qixue* volume of the *Gewu rumen*, Martin mentions the name "Ladena," which could have been Lardner. Ding Weiliang (W.A.P. Martin), *Gewu rumen*, 2:28b. Lardner was a prolific textbook writer who wrote extensively on topics related to the steam engine. For details about Lardner, see J. N. Hays, "Dionysius Lardner," *Oxford Dictionary of National Biography*, Volume 32 (Oxford: Oxford University Press, 2004), pp. 560–563.
49. "Preface," *Gewu rumen*, volume (Beijing: Tongwenguan, 1868), 1:1b.
50. Ding Weiliang, *Gewu rumen*, 5:1ab.
51. James Stormonth, *Etymological and Pronouncing Dictionary of The English Language* (Edinburgh and London: William Blackwood and Sons, 1881), p. 179.
52. Ding Weiliang, *Gewu rumen*, 2:29a.
53. The *Gewu rumen* uses the term *tanjin*, which literarily means "the pure form of carbon," to translate the element of carbon. However, it followed the *Bowu xinbian* in using the term *yangqi* ("life-supporting air") to translate "oxygen." Ding Weiliang, *Gewu rumen*, 3:2ab.
54. Ding Weilian, *Gewu rumen*, 3:2b.
55. The *Reqi* chapter uses daily-life experiences to explain why there is no such thing as cold. For example, it explains that the water in a well feels warm in winter but cold in summer because water retains heat better than air. Ding Weiliang, *Gewu rumen*, 3:2b–3a.
56. After these discussions, the *Reqi* chapter gives descriptions of thermal contraction and expansion. It tells its readers that when building bridges or boats with iron, one should consider the extent of expansion and contrac-

tion and leave gaps between sections to prevent the building materials from becoming distorted. Ding Weiliang, *Gewu rumen*, 3:3a.

57. The text discusses how thermal expansion and contraction might affect the accuracy of a pendulum clock and suggests sandwiching an iron rod between two lead rods to make the pendulum rod. The other solution is to use a hollow pendulum rod and fill it with mercury. Ding Weiliang, *Gewu rumen*, 3:3a–4a.

58. The text describes the experiment of inserting a mercury-filled glass tube into ice water and marking the position of the mercury as 32 degrees, then putting the glass tube into boiling water and marking the mercury level as 220 degrees, then dividing the two marks into 180 equal segments. That is the Fahrenheit thermometer used in Britain and the United States. However, because mercury freezes at temperatures below minus 40 degrees, alcohol can be used to produce the thermometer. The text also gives descriptions of different types of thermometers including one that uses alcohol instead of mercury. Ding Weiliang, *Gewu rumen*, 3:4b–8a.

59. One of the proponents of such a view was the British chemist John Dalton (1766–1844), who invented the chemical atomic theory. Hasok Chang, *Inventing Temperature*, p. 66.

60. The Scottish chemist William Irvine (1743–1787), who studied under Joseph Black and learned about latent heat, postulated that latent heat phenomena were consequences of changes in bodies' capacities to absorb caloric. For more details, see Hasok Chang, *Inventing Temperature*, pp. 168–170.

61. Ding Weiliang, *Gewu rumen*, 3:16b–17b.
62. Ding Weiliang, *Gewu rumen*, 3:18a–19d.
63. Ding Weiliang, *Gewu rumen*, 3:12ab.
64. Ding Weiliang, *Gewu rumen*, 3:14b–15a.
65. Feng Shanshan, *Jindai xifang rexue zai Zongguo de chuanbo*, PhD thesis, University of Inner Mongolia (2019), p. 45.
66. Ding Weiliang, *Gewu rumen*, 3:21b–22a.
67. Ding Weiliang, *Gewu rumen*, 3:23a.
68. For details about the wave theory of heat, which was discussed among physicists between the 1830s and the 1850s, see Stephen G. Brush, 'The Wave Theory of Heat: A Forgotten Stage in the Transition from the Caloric Theory to Thermodynamics', *The British Journal for the History of Science* 5:18 (1970), pp. 145–167.
69. Dionysisu Lardner, *Handbook of Natural Philosophy and Astronomy*, Second Course, Heat, Magnetism, common Electricity, voltaic Electricity (Philadelphia: Blachard and Lea, 1853), p. 179.
70. For more details about science teaching at the *Tongwen guan*, see David Wright, *Translating Science*, pp. 296–307.

71. Benjamin A. Elman, *On Their Own Terms*, p. 321.
72. Xiong Yuezhi, *Xixue dongjian*, p. 298.
73. Yong Mei, Feng Lisheng, "*Gewu rumen* zai Riben de liubo, *Xibei daxue xuebao* (*Journal of Northwestern University*, Natural Science Edition), Number 2, (2013), pp. 157–162.
74. Liu Zhixue, Chen Yunben, *Ziran bianzhengfa tongxun* (*Journal of Dialectics of Nature*), 40:5 (2018), p. 81.
75. Feng Shanshan, "Jindai xifang rexue zai Zhongguo de chuanbo," p. 44.

References

Amelung, Iwo, "Some Notes on Translations of the *Physics Primer* and Physical Terminology in Late Imperial China," *Wakumon* 或問 11:8 (2004), pp. 11–33.

Arnott, Neil, *Elements of Physics or Natural Philosophy* (London: Longman, Rees, Orme, Brown, and Green, 1833).

Bayliss, Robert A., and Ellis, C. William, "Neil Arnott, F.R.S., Reformer, Innovator and Popularizer of Science, 1788–1874," *Notes and Records of the Royal Society of London* 36:1 (1981), pp. 103–123.

Biggerstaff, Knight, *The Earliest Modern Government Schools in China* (Ithaca, NY: Cornell University Press, 1961).

Brush, Stephen G, "The Wave Theory of Heat: A Forgotten Stage in the Transition from the Caloric Theory to Thermodynamics," *The British Journal for the History of Science* 5:18 (1970), pp. 145–167.

Bryant, Lynwood, "The Role of Thermodynamics in the Evolution of Heat Engines" *Technology and Culture* 14:2 Part 1 (1973), pp. 152–165.

Cardwell, D.S.L., *From Watt to Clausius: The Rise of Thermodynamics in the Early Industrial Age* (London: Heinemann, 1971).

Carnot, Sadi, *Reflections on the Motive Power of Heat*, edited and translated by Robert Henry Thurston (New York: J. Wiley & Sons, 1890).

Chang, Hasok, "Thermal Physics and Thermodynamics," in Jed Buchwald and Robert Fox (eds), *The Oxford Handbook of the History of Physics* (Oxford: Oxford University Press, 2013), pp. 473–507.

Chang, Hasok, *Inventing Temperature: Measurement and Scientific Progress* (Oxford: Oxford University Press, 2004).

Chan Man Sing, "Sinicizing Western Science: The Case of *Quanti xinlun*," *T'oung Pao* 98 (2012), pp. 528–556.

Chen Wancheng 陳萬成 (Chan Man Sing), Luo Wanwei 羅婉薇 (Law Yuen Mei), Kuang Yongheng 鄺詠衡 (Kwong Wing Han), "Wan Qing xiyixue de yishu: yi *Xiyi luelun, Fuying xinshuo* liang ge gaoben weili 晚清西醫學的譯述：以《西醫略論》,《婦嬰新說》兩個稿本為例," *Journal of Chinese Studies, The Chinese University of Hong Kong* 56 (2013), pp. 243–292.

Conant, James Bryant, "The Overthrow of the Phlogiston Theory: The Chemical Revolution of 1775–1789," in James Bryant Conant (ed), *Harvard Case Histories in Experimental Science*, volume 1 (Cambridge, MA: Harvard University Press, 1966), pp. 65–116.

Covel, Ralph R., *W.A.P. Martin, Pioneer of Progress in China* (Washington, DC: Christian University Press, 1978).

Cullen, Christopher, "The Science/Technology Interface in Seventeenth-Century China: Song Yingxing on '*qi*' and the '*wu xing*'," *Bulletin of the School of Oriental and African Studies, University of London* 53:2 (1990), pp. 295–318.

Dai Nianzu 戴念祖 (ed.), *Zhongguo kexue jishu shi: Wulixue juan* 中国科学技术史: 物理学卷 (Beijing: Kexue chubanshe, 2001).

Dickinson, H.W., "The Steam-Engine to 1830," in Charles Singer, E.J. Holmyard, A.R. Hall, and Trevor L. Williams (eds), *A History of Technology*, Volume 4: The Industrial Revolution, c1750 to c1850 (Oxford: Oxford University, 1958), pp. 168–198.

Ding Weiliang (W.A.P. Martin), *Gewu rumen* (Beijing: Tongwenguan, 1868).

Duus, Peter, "Science and Salvation in China: The Life and Work of W.A.P. Martin (1827–1916)," in Kwang-ching Liu (ed.), *American Missionaries in China: Papers from Harvard Seminars* (Cambridge, MA: Harvard East Asian Research Center, 1970), pp. 11–41.

Elman, Benjamin, *On Their Own Terms: Science in China, 1550–1900* (Cambridge, MA: Harvard University Press, 2005).

Feng Shanshan 冯姗姗, "Jindai xifang renxue zai Zhongguo de chuanbo 近代西方热学在中国的传播," PhD Thesis, Inner Mongolia 2019.

Feyerabend, Paul, "Explanation, Reduction and Empiricism," in H. Feigl and G. Maxwell (ed.), *Scientific Explanation, Space, and Time*, Minnesota Studies in the Philosophy of Science, volume 3 (Minneapolis: University of Minneapolis Press, 1962), pp. 28–97.

Fox, Robert, "Dalton's Caloric Theory," in D.S. Cardwell (ed.), *John Dalton and the Progress of Science* (Manchester: Manchester University Press, 1968), pp. 109–124.

Fu Deyuan 傅德元, *Ding Weiliang yu jindai Zhong Xi wenhua jiaoliu* 丁韙良與近代中西文化交流 (Taipei: Taiwan University Press, 2013).

Gregory, Joshua C., *Combustion from Heracleitos to Lavoisier* (London: Edward Arnold & Co., 1934).

Hays, J.N., "Dionysius Lardner" *Oxford Dictionary of National Biography*, Volume 32 (Oxford: Oxford University Press, 2004), pp. 560–563.

He Xin 合信 (Benjamin Hobson), *Bowu xinbian* 博物新編, the 1855 Shanghai edition kept in the Kuo Ting-yee Library, Academia Sinica.

Henderson, John B., *The Development and Decline of Chinese Cosmology* (New York: Columbia University Press, 1984).

Hills, Richard L., *Power from Steam: A History of the Stationary Steam Engine* (Cambridge: Cambridge University Press, 1989).
Kim, Yung Sik, *The Natural Philosophy of Chu His (1130–1200)* (Philadelphia: American Philosophical Society, 2000).
Kuhn, Thomas, *The Structure of Scientific Revolution* (Chicago: Chicago University Press, 1962).
Lardner, Dionysisu, *Handbook of Natural Philosophy and Astronomy*, Second Course, Heat, Magnetism, Common Electricity, Voltaic Electricity (Philadelphia: Blachard and Lea, 1853).
Li Jianmin 李建民, "*Bencao gangmu* Huo bu kaoshi 本草綱目火部考略," *Bulletin of the Institute of History and Philology, Academia Sinica* 73:3 (2002), pp. 395–441.
Liu An 劉安, *Huainan honglie jijie* 淮南鴻烈集解, edited by Liu Wendian 劉文典 (Beijing: Zhonghua, 1989).
Liu Zhixue 刘志学, Chen Yunben 陈云奔, "Ding Weiliang bianyi wuli jiaokeshu pingxi 丁韪良编译物理教科书评析," *Ziran bianzhengfa tongxun* 自然辩证法通讯, 40:5 (2018), pp. 80–87.
Liu, Lydia H. (ed.), *Tokens of Exchange: The Problem of Translation in Global Circulations* (Durham: Duke University Press, 1999).
Liu, Lydia H., "Translating International Law," in idem., *Clash of Empires: The Invention of China in Modern World Making* (Cambridge, MA: Harvard University Press, 2006), pp. 108–139.
Lloyd, G.E.R., *Ancient Worlds, Modern Reflections: Philosophical Perspectives on Greek and Chinese Science and Culture* (New York: Oxford University Press, 2004).
Martin, W.A.P., *A Cycle of Cathay or, China, South and North with Personal Reminiscences* (New York: Fleming H. Revell Company, 1900).
Mendosa, E., "A Sketch for a History of Early Thermodynamics," in Spencer Weart and Melba Phillips (eds.), *History of Physics* (New York: American Institute of Physics, 1985), pp. 18–24.
Morris, Robert J., "Lavoisier and the Caloric Theory," *The British Journal for the History of Science* 6:1 (1972), pp. 1–38.
Needham, Joseph, *Science and Civilisation in China*, Volume 4: Physics and Physical Technology, Part 1: Physics (Cambridge: Cambridge University Press, 1962).
Onozawa Seiichi 小野泽精一, Fukunaga Mitsuji 福永光司, Yamanoi Yū 山井涌 (eds.), *Qi de xixiang: Zhongguo ziranguan yu ren de guannian de fazhan* 气的思想: 中国自然观与人的观念的发展, translated by Li Qing 李庆 (Shanghai: Shanghai renmin chubanshe, 1990).
Partington, J.R., and MacKie, D., *Historical Studies on the Phlogiston Theory* (New York: Arno Press, 1981).

Roller, Duane, "The Early Development of the Concepts of Temperature and Heat: The Rise and Decline of the Caloric Theory," in James Bryant Conant (ed.), *Harvard Case Histories in the Experimental Sciences* (Cambridge, MA: Harvard University Press, 1957), volume 1, pp. 117–214.

Shen Guowei 沈國威, Uchida Keiichi (內田慶市), Matsuura Akira (松浦章) (eds.), *Xia'er guanzhen, fu jieti, suoyin* 遐爾貫珍, 附解題、索引 (Shanghai: Cishu chubanshe, 2005).

Sivin, Nathan, *Traditional Chinese Medicine in Contemporary China* (Ann Arbor: University of Michigan Centre for Chinese Studies, 1987).

Smith, Crosbie, *The Science of Energy: A Cultural History of Energy Physics in Victorian Britain* (Chicago: Chicago University Press, 1998).

Song Yingxing 宋應星, *Ye yi; Lun qi; Tan tian; Si lian shi* 野議; 論氣; 談天; 思憐詩 (Shanghai: Shanghai renmin chubanshe, 1976, reprint of seventeenth century editions).

Stormonth, James, *Etymological and Pronouncing Dictionary of The English Language* (Edinburgh and London: William Blackwood and Sons, 1881).

Thomson, William, "On the Dynamical Theory of Heat," *Transactions of the Royal Society of Edinburgh* 21, part 1, (1853), pp. 261–288.

Wang Yangzhong 王扬宗, "Hexuli *Kexue daolun* de liangge Zhongyi ben: Jiantan Qingmo kexue yizhu de zhunquexing 赫胥黎'科学导论'的两个中译本: 兼谈清末科学译著的准确性," *Zhongguo keji shiliao* 中国科技史料 21:3 (2000), pp. 207–221.

Wright, David, *Translating Science: The Transformation of Western Chemistry into Late Imperial China, 1840–1900* (Leiden: Brill, 2000).

Xi Zezong 席泽宗 (ed.), *Zhongguo kexue jishu shi: kexue sixiang juan* 中国科学技术史: 科学思想卷 (Beijing: Kexue chubanshe, 2001).

Xiong Yuezhi 熊月之, *Xixue dongjian yu wan Qing shehui* 西学东渐与晚清社会, revised edition (Beijing: Zhongguo renmin daxue chubanshe, 2011).

Yong Mei 咏梅, Feng Lisheng, "*Gewu rumen* zai Riben de liubo 格物入门在日本的流播," *Xibei daxue xuebao* 西北大学学报 (Natural Science Edition), 2, (2013), pp. 157–162.

CHAPTER 4

Achievements and Constraints of Late Qing Translations of Heat, 1868–1895

By the 1850s and 1860s, books like the *Bowu xinbian* and the *Gewu rumen* gave their Chinese readers an introduction to the basics of Western science. They might not have been a "linguistic turn" in a Kuhnian sense in the history of science in China, but they did mark the beginning of a large number of science and technology publications, in which new terms and new technical language were introduced to convey new theories that could not have been expressed by traditional Chinese natural philosophy. After the 1870s, various books and magazine articles, which were mostly translated or authored by foreign missionaries in collaboration with their Chinese colleagues, were published. The subjects, ranging from chemistry, mathematics, astronomy, and physics to mining, military strategy, and international politics, formed a body of knowledge called *Xixue* (Western learning). As shown by Xiong Yuezhi and many others, Western learning attracted the Chinese literati, especially those who had opportunities to witness Western technological power in the treaty ports of Shanghai, Guangzhou, or Tianjin.[1] How might those translated texts have made an impact on the Chinese understanding of nature and technology?

This chapter examines the translated texts on heat and the steam engine published between the 1870s and the 1890s. Specifically, it focuses on how the elements of thermodynamics could be translated into Chinese and absorbed by the Chinese literati. It also reflects on how those publications might have enhanced China's transfer of steam engine technology.

Doubting *QI*

Xu Shou and Hua Hengfang, the two Chinese literati who experimented on the steam engine and constructed China's first domestically-built steamer, the *Huanghu*, were traditionally educated but were not successful in imperial examinations.[2] While Hua was interested in mathematics, Xu focused on optics. We do not know much about Hua's thinking in natural philosophy, but according to a biography, Xu had doubted traditional knowledge of horoscopes, geomancy (*fengshui*), sorcery, and prophecy, nor did he trust *li*, that is, the natural and moral principles of things, or *qi*, the origin of myriad things. Both conceptions were an essential part of neo-Confucianist cosmology.[3] Additionally, Xu was interested in making both musical and scientific instruments. He had done an experiment on the thermometer, observing the phenomena of thermal expansion and contraction in mercury. In an undated letter to Hua, he wrote:

> I have doubts in [my understanding of] the properties of mercury in your thermometer (*hanshu biao*) because the experiments have not gone well. Today I have carried out experiments based upon the method we discussed previously, and I find that its nature is [such that] it is able to expand and contract. It was a quick matter to establish [its behavior] [whether by] using fire to heat it [or] using my breath to warm it. I then put it outside, to test the warmth and cold of nature, and everything also went well. On the 25th and 26th of the first lunar month, the weather was fine, and I put it in the open air early in the morning [and then] put it ins the midday sun, and estimated [the difference] as 5 degrees, and found I could judge it to [within] one per cent. [Even if the thermometer] were made ingeniously, this [amount of inaccuracy] would not be a matter for regret. Regarding the nature of the expansion and contraction of mercury, have we not looked at each other and laughed at the way in which, in the past, because we did not understand it, we used *li* and *qi* to discuss this matter, seeking profundity through suchlike superficialities! However, when I used the fire to heat it, another doubt arose. Having made a mark on the thermometer where [the mercury level] had reached under the sun, I then brought the thermometer near the fire, and, contrary [to my expectations], the mercury contracted: thus it would appear that the nature [of mercury] with respect to cold and warmth must be subdivided into [two different natures, with respectively] a gradual and a rapid aspect, and that its nature [when subjected to] rapid heating is opposite to that [when subjected] to gradual heating. I have still not fathomed the principle [governing this]. What do you make of it?[4]

In other words, by experimentation, albeit a crude one, Xu learned how the mercury thermometer could give readings of the level of "the warmth and cold of nature" on the basis of thermal expansion and contraction. He had felt that traditional Chinese natural philosophy could not give him a satisfactory explanation for the effect of heat on mercury's volume. However, he was puzzled by the fall of the mercury when he heated the glass bulb with a naked flame. He suspected that it might have something to do with the sudden heating of the instrument and that a well-made thermometer would probably produce satisfactory experimental results. Nevertheless, he still speculated that mercury could have different expansion rates when it was heated in different ways. David Wright has suggested that what Xu observed probably resulted from a poorly made thermometer whose glass expanded faster than the mercury.[5]

This letter was probably written before Xu and Hua were invited into the service of the private secretariat of Zeng Guofan, then the governor-general of Jiansu, Jiangxi, and Anhui. They read the *Bowu xinbian* and started to work on their steam engine in 1861. Xu's experience with the thermometer and his work on the steam engine could have facilitated deeper thinking into the role of heat in the steam engine. Unfortunately, they did not leave any documents revealing what they thought about heat toward the completion of the *Huanghu* in 1865. Because of that experience, they could have become pioneering experimental scientists in nineteenth-century China. Nevertheless, regardless of the intellectual foundation, the politico-social situation of China in the 1860s would not have allowed Chinese experimentalists to develop a new set of technical language to replace "*qi*" with "energy" at their own pace because the humiliation of the Second Opium War and the devastation of the Taiping Rebellion had forced the Qing government to adopt the Western technologies of steam warships and firearms hurriedly. Besides, it was not possible for Xu and Hua to carry on as independent scientists because they did not have the socioeconomic means to do so. As part of Zeng Guofan's private staff, they were transferred to the newly established Jiangnan Arsenal as translators of Western science and technology treatises. Notably, Xu's second son Xu Jianyi, who had learned Western science from his father and also became one of the members of Zeng Guofan's staff about Western science and technology, also joined the Jiangnan Arsenal translation project.

Translating Textbooks of Engine-room Management

In 1867, when Li Hongzhang, then governor of Jiangsu province, established the Jiangnan Arsenal in Shanghai, his aim was to use the formerly foreign-owned machine shop to manufacture firearms and ammunition to supply Chinese troops. However, before he could further develop the Arsenal's firearm production capacity, he was appointed governor-general of Hunan and Hubei. In 1870, his patron Zeng Guofan, formerly the governor-general of Zhili, resumed the position of governor-general of Jiangsu, Jiangxi, and Anhui. Committed to domestic steamship building, Zeng initiated the Arsenal's shipbuilding projects, and a wooden-hulled steamboat fitted with a refurbished engine was built in 1868.[6]

Furthermore, Zeng Guofan knew that developing modern technology required science and mathematics, and hence earned permission from the imperial court to establish the Translation Department (*Fanyi guan*) under the Arsenal, aiming to translate books that benefitted China's technological development.[7] Apart from Xu Shou and Hua Hengfang, he also recruited foreigners, such as the British missionary John Fryer (1839–1928) and Alexander Wylie (1815–1887), who had learned Chinese and were slightly better informed in Western science and technology than most Chinese literati.[8] The foreign translators might have their own preferences in the selection of source texts, but Jiangnan Arsenal officials, who considered the technologies of steam warship building and firearm manufacturing essential to China's needs, would have made the final decisions. On several occasions, senior officials such as Zeng Guofan and Li Hongzhang would demand books on certain subjects to be translated.[9] Soon, the translators started to produce texts whose subjects ranged from mathematics, applied chemistry, engine room manuals, mechanical engineering, mining, electricity, and medicine to military strategy and law.

John Fryer's career in China is well-documented, but a few details are worth mentioning here.[10] Trained as a school teacher in England, he was recruited by the London Missionary Society to teach in Hong Kong in 1861, where he learned Chinese. In 1863, he was appointed as an English professor at the School of Combined Learning in Beijing. In 1865, he left this job and set up a small Anglo-Chinese school in Shanghai. At the same time, he became the editor of the Chinese newspaper *Shanghai Xinbao* (*Shanghai Gazette*). It was then that Fryer became interested in showing scientific experiments to his Chinese students.

Alexander Wylie (1815–1887) was trained as a cabinet maker after completing school in London. During his apprenticeship, he taught himself Chinese by reading a Chinese grammar book in Latin. In 1846, he was recruited by the London Missionary Society to superintend the society's press (*Mohai shuguan* or Inkstone Press) in Shanghai. Wylie started translating parts of the New Testament into Chinese and, at the same time, editing the Chinese magazine, *Liuhe congtan* (*Shanghae Serial*), which offered news, commercial intelligence, and popular science essays to Chinese readers.[11] He made friends with reform-minded Chinese literati such as Xu Shou and Li Shanlan, a prominent Chinese mathematician. Wylie also studied Chinese culture and traditional science and, in 1853, with help from Li Shanlan, published a paper entitled "Jottings on the Sciences of the Chinese Arithmetic" in the *North China Herald*, in which he traced the history of Chinese mathematics and used Western mathematical notations to translate Chinese equations. He also worked with Li Shanlan to translate mathematical treatises on elementary mathematics, analytical geometry, and calculus into Chinese. Notably, they retained Western mathematical notations but replaced Arabic numerals with Chinese characters.[12] Hence, although neither Wylie nor Fryer were trained as scientists or engineers, at the onset of the Jiangnan Arsenal's translation project, Wylie was more learned than Fryer in both Chinese and Western mathematics and sciences, and had more experience in translation.

Chapter 3 has already discussed the process of translation at the Jiangnan Arsenal according to John Fryer's account. Here, we need to reiterate the issues involved in collaborative translation. Because foreign translators might have learned Chinese as adults, their writings were usually clumsy and not publishable. However, Chinese translators did not read European languages, nor were they necessarily familiar with the contents of the source texts. As such, foreign translators would first have to understand the source texts and then orally explain their contents to the Chinese translators, who would have to retranslate these ideas into proper, publishable classical Chinese. Had there been errors, Fryer stated in 1890, "[t]he responsibility for whatever undue haste or carelessness may characterize my work rest rather on my Chinese colleague than myself."[13] David Wright has noted that Fryer was shifting the blame to his Chinese colleagues. Because no foreign translators were professionally trained as scientists or engineers, even if they were eager to acquire new knowledge, they often misunderstood new sciences and refused to change despite the rapid, radical changes of science and technology at the time, causing confusion in terminology.[14] However, the point here is not to blame either the foreign or Chinese translators for their errors or inability to catch up with the latest scientific theories or state-of-the-art technology but rather to stress the difficulties in introducing Western science into China in the second half of the nineteenth century.

Soon after he joined the Jiangnan Arsenal's translation project, Xu Shou started to work with John Fryer on the chemistry textbook *Huaxue jianyuan* (*Mirroring the Origins of Chemistry*), which laid down some of the principles of translating chemical nomenclature into Chinese. The book was published in 1871.[15] By that time, Fryer had also worked with other Chinese translators on coal mining and gunpowder making.[16] In other words, the translators at the Jiangnan Arsenal were working on different subjects at the same time. It would have been difficult for them to delve into all the subjects they were translating.

During the same period of time, Wylie worked with Xu Shou on the translation of the *Qiji faren* (*A Primer on the Steam Engine*), which was printed by the Jiangnan Arsenal in 1871. Their source text was the third edition (1855) of *The Marine Steam Engine* by Thomas J. Main and Thomas Brown. Main was a mathematical professor at the Royal Naval College, and Brown was the Chief Engineer of the Royal Navy. According to the book's preface, the authors felt that most contemporary textbooks on the steam engine lacked practical knowledge about engine management, and hence, they, on the basis of their experiences in instructing naval officers, intended to offer insights about steam as applied to nautical purposes. Their intended readers were naval officers who would be managing marine engines on board royal or merchant naval fleets.[17]

The Marine Steam Engine's introductory chapter states that the book does not enter into the discussion on the nature of heat but rather focuses on the effects of heat on volume, especially in the case of steam. After introducing different types of thermometers, and the calorimeter, which is an instrument that measures heat capacity (we will discuss more later), it then discusses the phenomena of heat conduction, convection, and radiation. It also gives descriptions of concepts including capacity for heat, latent heat, sources of heat, effects of pressure on boiling points, and steam pressure. Building on these basic concepts, the book goes on to discuss technical matters of the marine steam engine. Chapter 2 gives detailed descriptions of the commonly installed tubular square or cylindrical boilers. Chapter 3 outlines the steam engine's history including the Newcomen engine, the Watt engine, and the latest marine engines, and

Chap. 4 discusses the motions of engine parts such as the piston and the crank. Because calculations of temperature and pressure are required to maintain efficiency and to prevent serious accidents such as explosions, this chapter contains trigonometric and calculus formulae for calculating motions and work. Subsequently, Chap. 5 provides instructions on practical engine operations such as filling up the boiler, firing, and monitoring the steam gauge and the valves. Chapters 6, 7, and 8 discuss various practical problems a steamship might face, from gear damage to boiler explosions. Finally, Chap. 9 details how to measure the efficiency of the marine engine by duty, which the book defines as the consumption of fuel divided by effective horsepower.

In short, *The Marine Steam Engine* focuses on the practical technicalities of engine management, not a theoretical understanding of heat and the engine. It asserts that engine efficiency can be achieved by careful management of temperature and pressure and can be measured by calculating duty. This idea had been prevalent since James Watt; the only difference was the complexity of the engine and the mathematical skills, especially algebra, logarithms, trigonometry, and some calculus, that it required readers to have.

The Jiangnan Arsenal's selection of such a source text must have something to do with its officials' intention to provide reference books to future Chinese naval forces, which were in the process of forming at that stage. (Chapter 5 will discuss this issue further.) Xu Shou already had experience in building marine steam engines, and Wylie's knowledge of mathematics would have been helpful in their translation work. Indeed, they made great efforts to follow the sentences as closely as possible and reprinted all the original illustrations.

The new terms used to introduce new ideas deserve our attention. To translate "steam," Wylie and Xu used the character "*qi*," which has the radical of water but not the character that was often associated with air or the metaphysical *qi*. Using the new character would avoid confusion between "steam" and the metaphysical *qi*. Later translations of "steam" would continue this usage. To translate "caloric," Xu and Wylie used the term "*reyuan*" (literally "the element of heat"), which is quite fitting. Their use of "*yinre*" ("hidden heat") to denote latent heat is also suitable. Indeed, this was the first book that introduced the concept of latent heat to Chinese readers.[18] It was also the first mention of the idea of the calorimeter (*liangreqi*, literally "heat measuring device"). "Capacity for heat" was translated as "*rongrelü*," which literally means the "ratio of containing heat." The ideas of "duty" and "horsepower" were also new. Wylie and Xu chose "*gonglü*" and "*mali*," respectively, to translate them. Both terms survive to the modern day, although the term "duty" has been replaced by "power" in modern English physics terminology.[19] Despite these innovations, Wylie and Xu's translation of "efficiency" with the term "*zhiyong*" is slightly awkward. *Zhiyong* literally means "make full use of," which is close but not quite the meaning of "efficiency."

Furthermore, although Xu Shou was not a mathematician, Wylie was experienced in dealing with mathematical texts. They translated the mathematical equations by replacing the Arabic numerals with Chinese characters and the Latin or Greek alphabetic with heavenly stems (*tiangan*) and earthly roots (*dizhi*), the traditional Chinese counting system. However, they preserved the mathematical symbols for the four basic operations: the radical sign, the fraction bar, the bracket, and the logarithm.[20] The nineteenth-century Chinese literati would probably have found the *Qiji faren* rather technical. To understand the contents, they would have to read the *Gewu rumen*, which provided a basic knowledge of mechanics, steam pressure, and the mechanism of the steam engine, beforehand. Otherwise, they would have found the technical details of the *Qiji faren* inaccessible. To understand its mathematical equations, readers would also have to read Wylie's works on algebra, logarithms, trigonometry, and calculus.[21]

The Jiangnan Arsenal carried on publishing more engine-room manuals, but none of them matched the level of technical knowledge of the *Qiji faren*. In 1872, the Arsenal published the *Qiji biyi* (*A Manual of the Steam Engine*), which was translated by John Fryer and Xu Jianyin from the eleventh edition of John Bourne's *A Catechism of the Steam Engine: A Handbook of the Steam Engine*, a popular textbook first published in London in 1847. A year later, in 1873, Fryer again worked with Xu Jianyin, publishing the *Qiji xinzhi* (*Newly-Designed Steam Engine*) by translating the first edition of Nicholas Procter Burgh's *Pocket Book of Practical Rules for the Proportions of Modern Engines and Boilers for Land and Marine Purposes*, which was published in London in 1864. Both books are basic engine room manuals that provide practical operational knowledge without touching on theoretical issues. Managing pressure, temperature, and force is their major concern.[22]

Although managing pressure, temperature, and force were indeed practical issues in operating the steam engine, the scientific principles of these books appeared to be outdated in the 1870s. From the 1850s, professional science and technology journals had been publishing engineering

discussions on how one might apply the new science of energy to engine designs. New textbooks of thermodynamics were written to instruct younger generations of engineers. Most famously, *A Manual of the Steam Engine and Other Prime Movers* (1859) by the British engineer and physicist William Rankine, *Die mechanische Wärmetheorie* (*The Mechanical Theory of Heat*, 1864) by the German physicist Rudolf Clausius, as well as *Treatise on Natural Philosophy* (1867) by William Thomson and Peter Guthrie Tait were published and translated into major European languages. All of these books were theoretically and mathematically demanding. Representing the forefront of physics and engineering, they were widely read and discussed among physicists and engineers in Europe and the United States. The impact of these works was profound. To increase efficiency, engineers started to experiment with different working fluids to replace steam. Their efforts ultimately led to the invention of the internal combustion engine.[23]

One might argue that modern science and technology in China in the 1870s was still in its infancy, so it would have been unrealistic to expect that translating such theoretically demanding textbooks would have helped China develop steam engine technology at this stage. Actually, if these theoretical textbooks could not help, neither could textbooks on engine-room management. It is for exactly this reason that one has to look at the institutional settings for transferring, storing, and developing new knowledge in China.

The Qing government's self-strengthening initiatives, which began in 1865, included establishing new schools to increase young Chinese or Bannermen's foreign language skills and knowledge about Western powers to improve the government's ability to deal with foreign affairs.[24] Senior officials such as Prince Gong and Li Hongzhang knew that science and mathematics were important in the Qing government's efforts to import steamship and firearm technologies. However, they did not seem to fully understand the depth and complexity of science and mathematics required by nineteenth-century technology. Beijing's School of Combined learning (*Tongwen guan*) only taught foreign languages when it was established in 1862 and did not add courses on mathematics and astronomy into its curriculum until 1866. As discussed in Chap. 3, it only taught primary-school-level physics and mathematics using textbooks, such as the *Gewu rumen*, written or translated by its own teachers. In Shanghai, Li Hongzhang established a similar language school, the *Guangfangyan guan*, in 1863. Li also suggested that a similar school be established in

Guangzhou, named the *Tongwen guan*. In 1869, the Shanghai school was merged into the Jiangnan Arsenal, and new courses in mathematics, marine engineering, and shipbuilding were included in its curriculum.[25]

None of the Jiangnan Arsenal publications on the steam engine were used in any of the courses in these schools. According to Fryer's reminiscence in 1880, the teachers of shipbuilding and marine engineering at the Jiangnan Arsenal's school did not use the translation department's publications because foreign teachers did not know Chinese.[26] China had only one engineering school, the School of Naval Construction in the Fuzhou Navy Yard, that taught marine engineering and naval construction from their scientific and mathematical principles to their practices. This school recruited French teachers and engineers to teach Chinese students with French textbooks. (Chapter 5 will discuss this school in more detail.) Though the Jiangnan Arsenal's own school might have been teaching students some scientific subjects, they did not train engineers.

Nevertheless, the Chinese literati did read the Jiangnan Arsenal's publications. In 1898, Huang Qingcheng, a literatus who had learned mathematics, commented that the *Qiji faren* was easier to understand than the other Jiangnan Arsenal translations.[27] We do not know on what grounds he made such an assessment. Nevertheless, Fryer and Huang's comments reveal one of the fundamental problems in transferring Western science into China. Although Chinese literati might have been interested in learning new knowledge, only a small number of them who had some knowledge of mathematics and physics would be able to understand these newly translated books. The discussion here does not intend to belittle the Jiangnan Arsenal Translation Department's efforts to introduce new knowledge into China; one certainly cannot ignore the impact of Fryer and Xu Shou's *Huaxue jianyuan* on Chinese chemical nomenclature. However, because those books were not translated in an organized way or used in any schools as part of the curriculum, whether one understood their contents had a limited impact on Chinese understandings of nature. The Chinese readership was simply unready for them.

The Jiangnan Arsenal continued to publish books on chemistry, astronomy, mathematics, physics, medicine, metallurgy, and mining.[28] However, it did not publish more books on the steam engine until 1894.[29] Probably aware of the limited readership for his translations at the Jiangnan Arsenal, John Fryer sought to produce popular science texts or primary-education textbooks, which would appeal to a wider readership.

Popular Science Texts: Edging Toward to a More Up-to-Date Theory

While working for the Jiangnan Arsenal, Fryer was also editing the popular science magazine *Gezhi huibian* (*The Chinese Scientific Magazine*), which he started in 1876. He gained support from the Society for the Diffusion of Useful Knowledge (*Guangxue hui*) in Beijing, which was closing down in 1875 and ending its monthly magazine *Zhongxi wenjian lu* (*The Peking Magazine*). Fryer had a Chinese literatus named Luan Xueqian help him with Chinese composition. Luan was from Shandong province and was educated at a missionary school in Dengzhou. The school taught elementary to secondary education level mathematics, chemistry, and physics. It also required students to attend experimental courses.[30] In other words, Luan and his classmates were among those who were better educated in science than the vast majority of Chinese literati.

In the *Gezhi huibian*'s first issue (second lunar month) in 1876, Fryer began to publish a series of essays collectively entitled "*Gezhi luelun*" ("Introduction to the Sciences"), which was a translation of an unknown children's textbook. In the series, he published a short essay on the steam engine, "*Qiji yaoshuo*" ("Basic Explanation of the Steam Engine"), giving a very simple description of how the expansive force of steam drove the steam engine.[31] In the fifth issue (sixth lunar month) of 1876, Fryer published an essay entitled "On Heat" (*Lun re*) as part of the *Gezhi luelun* series. The essay gives a very simple dynamical explanation of heat. It states that, instead of the earlier conviction that heat was an imponderable fluid, scientists have now been convinced that heat is the constant movement of the particles of matter: "when a body is heated, its particles are distant from each other… The hotter the body is, the farther the distance between its particles is." The essay then offers a short and simple discussion of heat diffusion, heat radiation, and the sun as the ultimate source of heat on earth.[32] Unfortunately, we have no information about how much Chinese literati had learned from these two simple essays.

The subjects published in the *Gezhi huibian* ranged from printing machines, gunpowder making, and sugar making to chemical knowledge, natural history, agriculture, and short biographies of scientists. The journal was highly popular: it was sold in Guangzhou, Ningbo, Tianjing, and many other treaty ports in China as well as the British colonies of Hong Kong and Singapore. It even reached the Japanese cities of Yokohama and Kobe. In its first year of publication, Fryer printed three thousand copies for every issue, which would be sold out within a year. In the following year, every issue would be sold out within days. In 1890, due to high demand, Fryer had to reprint every issue at least once.[33] Readers wrote letters to the editor asking questions about common natural phenomena or mathematics.[34] In other words, the *Gezhi huibian* had reached the general Chinese readership much better than the Jiangnan Arsenal's publications.

Confusing Energy with Force: The *Gezhi Qimeng*

Probably aware of Chinese readers' appetite for popular science and technology texts, in 1880, the Jiangnan Arsenal published the *Gezhi qimeng* (*Science Primers*), which included four books: *Huaxue* (chemistry), *Gewuxue* (physics), *Tianwenxue* (astronomy), and *Dilixue* (geography). These were translations from the *Science Primer* series, which were edited by Thomas Henry Huxley, a biologist and a strong advocate of Charles Darwin's theory of evolution; Balfour Stewart, a physicist; and Henry Roscoe, a chemist. The series was published in London beginning in 1872 with an introductory volume by Huxley.[35] Popular in the British and American textbook markets, the series continued to be revised and republished well into the early twentieth century.

The volume on physics was written by Balfour Stewart, who specialized in radiant heat and terrestrial magnetism. He was one of the British physicists who supported William Thomson's theory replacing the concept of force with energy in the 1850s.[36] In the preface, Stewart stated that the series intended to provide "the fundamental principles of science in a manner suited to pupils of an early age." Thus, he wrote *Physics* in simple, conversational prose that young children could relate to.[37] Energy was one of the core physical concepts in the book.

The Jiangnan Arsenal's volume of *Physics* was translated by the American missionary Young John Allen (1836–1907) and the Chinese translator Zheng Changyen. It was probably published in 1880. The same book was translated again by the British missionary Joseph Edkins (1823–1905) and published in 1886 by the Chinese Imperial Maritime Customs.[38]

Allen taught himself Chinese after he arrived at Shanghai in 1860. In 1864, he was recruited by the School for the Diffusion of Languages (*Guangfangyan guan*) in Shanghai as an English teacher. He also joined the Jiangnan Arsenal as a translator. From 1868, he became the editor of the Chinese weekly journal *Jiaohui xinbao* (*Church News*), which not only reported church news in

China but also published popular science and technology essays. The journal was later renamed *Wanguo gongbao* (*The Chinese Globe Magazine*) in 1874. Unfortunately, I know nothing about Zheng Changyen.

Joseph Edkins was active in translating Western science into Chinese after he was sent by the London Missionary Society to Shanghai in 1848. He became Alexander Wylie's colleague at the London Missionary Society Press and learned Chinese. From 1852 to 1858, he edited the annual journal *Huayang heh tongshu* (*Chinese and Foreign Concord Almanac*). In 1859, he worked with the mathematician Li Shanlan on the translation of *Zhongxue* (*Mechanics*) from the British polymath William Whewell's *An Elementary Treatise on Mechanics*, which was first published in London in 1819. He also worked with a Chinese literatus on *Guanglun* (*On Light*), which was not printed until 1896.[39] Edkins was interested in studying the Chinese language and wrote extensively about the subject. He contributed many articles about science and technology to journals such as the *Gezhi huibian*.[40] In 1880, Edkins left the London Missionary Society to work for the Imperial Chinese Maritime Customs, where he continued to work on translations, among them, the *Science Primers* series.[41]

Science Primers: Physics is a book for children, hence, Stewart wrote it in an informal and nontechnical style. It includes ten sections: the forces of nature, the three states of matter and their properties, the motions of bodies, heated bodies, and electricity. The concept of energy appears in the "Moving Bodies" section. The book defines energy as the power of doing work, for example, raising a pound weight one foot high; one measures energy by measuring the amount of work that it can do. Although a moving body contains energy, a motionless body is also capable of doing work. For example, the fall of water from a high-level tank can drive a waterwheel. Balfour uses the analogy of money ready to be spent to explain the energy in a moving body and money in the bank to represent the energy in a motionless body.[42]

In the first chapter, which introduces the forces of nature, the *Primer* defines that force sets a body in motion or brings the body in motion to rest. Gravity is one of the chief forces of nature by which things are attracted to the earth. Hence everything falls down unless a force is applied in the opposite direction, forcing them up. In the "Heated Bodies" section, Stewart again uses conversational-style prose to describe a heated body as an "energetic body," by which he means a body that contains energy. Just as sound is a kind of vibratory motion, not matter, heat is really a kind of vibratory motion because when a body is heated, its tiny, unseen particles are jiggling. Then, on the foundation of that principle, the book explains thermal expansion and the mercury thermometer. The text also discusses specific heat, change of state, the latent heat of water and steam, boiling point, heat's inclination to diffusion, heat conduction and convection, and radiant heat.[43] In explaining the nature of heat, Stewart again uses the analogy of sound. He states that hitting a bell with a hammer transfers energy from the hammer to the bell, which vibrates and makes a sound. That is doing work. A blacksmith hitting a piece of lead makes a dull thud, but the energy of the blow causes the lead's particles to vibrate and is transformed into heat. Hence, energy is not force.[44]

Although it was a book for children, *Science Primers: Physics* would have been a good introduction to the concept of energy for Chinese readers in the 1880s. Allen and Zeng's prose is classical and succinct, unlike Stewart's informal style. It leaves out many sentences that adults might find unnecessary but which were purposely written to guide children's understanding.[45] In contrast, Edkins' prose is a literal translation of Stewart's text and hence long and slightly awkward. He probably did not have a Chinese collaborator, perhaps due to his confidence in his own Chinese composition or a lack of confidence in Chinese collaborators' ability to convey his ideas.[46]

Allen and Zeng's translation of Stewart's passage defining energy in repose uses the phase "*wu dong zhi li*" ("the force of [a] body's motion"), which reads more like force than energy. They state that moving bodies can do work (*zuo gong*) but still bodies also contain force (*li*). In comparison, Edkins uses the word "*li*" to denote energy. He also uses the phrase "the force to do work" (*zuo gong zhi li*) to define energy. The rest of the texts transmit Stewart's ideas quite accurately, but using the word *li* confuses the concept of energy with force. Chapter 3 has already discussed how the English word "energy" could mean "power" or "force," which could both be translated as *li*. As Iwo Amelung suggests, both Allen and Edkins's translations failed to give a terminological solution.[47]

More elementary-level textbooks were translated and published. As such, Fryer extended his translation career to textbook publishing by joining the School and Textbooks Series Committee (*Yizhi shuhui*), which was a press established in Shanghai by the General Conference of Protestant Missionaries in China in 1877. The committee's purpose was to publish both primary and secondary school textbooks in Chinese whose subjects ranged from mathematics, surveying, astronomy, geography, chemistry, geology, botany, zoology, anatomy, and physiology to history and philosophy. Fryer was appointed the editor of the committee.[48] He edited the *Gezhi tushuo* (*Illustrated Treatises on Science*) series, and in 1890, published a primary education textbook, *Rexue tushuo* (*Illustrated Treatise on Heat*). Unfortunately, we cannot find any information about the *Rexue tushuo*'s source text, nor do we know who helped Fryer polish his Chinese prose, although it could have been Luan Xueqian.

The *Rexue tushuo* provides more information than the *Gezhi qimeng*. It has two volumes. The first volume states that heat is the vibration of particles (*weidian, literally means minute particle*): the hotter the object, the faster the vibration. It also has the usual discussions of thermal expansion and contraction, heat transmission, convection, and radiation, as well as the effect of pressure on a fluid's boiling point. The second volume discusses specific heat, heat capacity, latent heat, the calorimeter, rarefaction of steam, and the relationship between light and heat.[49] The book also contains illustrations of experimental instruments, including the calorimeter. The discussion of the relationship between heat and light seems to suggest that Fryer was using a children's textbook that was written before the 1850s.

Because calorimetry is one of the key elements of thermal physics and for the benefit of the discussion below, we need to say a few words about the *Rexue tushuo*'s description of the calorimeter.[50] The text states that the purpose of the calorimeter (*liangre qi*, literally "heat measuring device") is to measure heat capacity (*rongre zhi lü*) (Fig. 4.1). It describes the calorimeter as consisting of three iron-sheet cup-shaped vessels of different sizes, one within another. Each of the two outer chambers has a funnel. To measure an object's heat capacity, one places an object into the inner chamber and fills the two outer chambers with crushed ice. The ice in the outer chamber absorbs the heat in the environment, and the ice in the second chamber absorbs the heat from the object in the inner chamber. Once the temperature of the object drops to zero degrees, one weighs and compares the differences between the water that drips from the two funnels.

It is also worth noting that in its discussion of Joule's experiment on the conversion between heat and work, *li* (force) is used to translate "work," as in the *Gewu rumen* discussed in the previous chapter. In the final few pages, the *Rexue tushuo* gives a description of the steam engine, suggesting that the power of the steam engine comes from heat. That is to way, this textbook is a proponent towards the Joule's mechanical equivalence of heat.

Iwo Amelung has already pointed out that the *Gezhi qimeng* sparked the Chinese literati's interest in physics.[51] The simple content and rich illustrations of the *Rexue tushuo* would have had a similar effect. All of

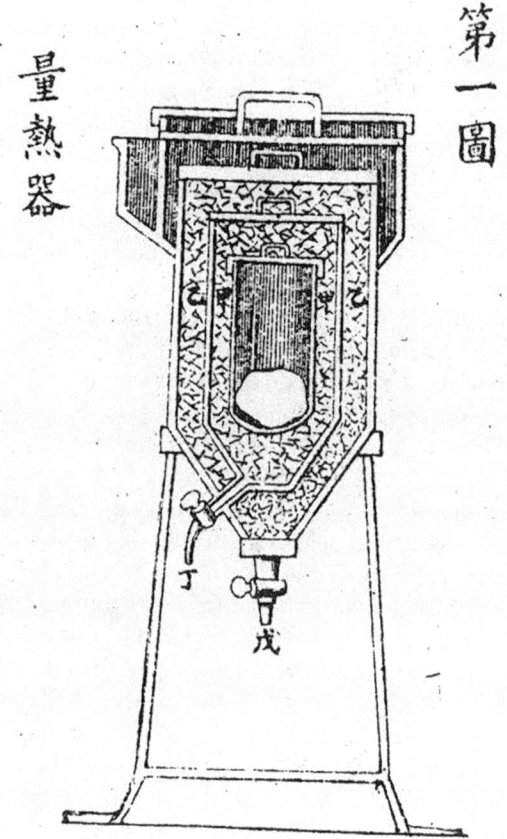

Fig. 4.1 The illustration of the calorimeter in the *Rexue tushuo*. (Source: Fu Lanya (John Fryer), *Rexue tushuo* 2:1a, (the 1890 edition kept in the Shanghai Library))

these publications remained at the elementary level. Their language is plain; their contents are nontechnical and contain no mathematical equations. As such, these works would have helped to diffuse a more up-to-date dynamical understanding of heat among their Chinese readers.

Moving Further Away from *Qi*: Chinese Writings About Heat

Once Xu Shou joined the Jiangnan Arsenal, he no longer conducted experiments but rather focused on translation works. He had to work on different subjects like chemistry, mining, and even autopsy at the same time. However, his knowledge of the steam engine was probably unmatchable among the vast majority of the Chinese literati in the 1870s. In the 1876 issue of the *Gezhi huibian*, Xu Shou published a short essay entitled *"Qiji minming shuo"* ("On Naming Steam Engine [Parts]") that gave a detailed description of a marine steam engine and the names of its parts. The essay describes how the boiler generates steam, which enters the cylinder through a valve that controls its direction. The expansive force (*zhangli*) of the steam forces the piston to move reciprocally. After doing work (*gong*), the steam enters the external condenser to be condensed by cold water, which creates a force of contraction (*suoli*). Xu states that the expansive force of steam is unlimited but the atmospheric force is fifteen pounds per square inch. Although Xu's description of the steam engine is more detailed than the *Bowu xinbian* or the *Gewu rumen*, it remains at a similar level theoretically.[52] Unfortunately, Xu Shou did not write anything further about the steam engine, so we cannot judge how his knowledge progressed along with his career as a translator. However, we know that he joined John Fryer in the new enterprise of promoting Western learning to the Chinese people.

When Fryer was editing the *Gezhi huibian*, he was also involved in running the *Gezhi shuyuan* (Academy for the Extension of Learning) or the Polytechnic Institution. This was a public reading room aimed at promoting knowledge regarding foreign countries and topics. It was originally conceived in 1873 by Walter Henry Medhurst, the British consul in Shanghai.[53] Medhurst proposed that the reading room should exhibit maps, scientific instruments, models of steam engines, telegraphic apparatus, and similar technologies. Additionally, the reading room should hold lectures in Chinese on scientific subjects of practical value. His proposal earned broad support from the foreign merchant community in Shanghai.[54] In 1874, a managing board that included Medhurst, Alexander Wylie, and John Fryer was formed.[55] In 1875, Xu Shou and his son Xu Jianyin were included on the board. Funds were raised from foreign and local Chinese mercantile communities as well as Qing government officials, especially Li Hongzhang. In 1876, the academy building was completed and opened to the public. In 1877, the *Gezhi shuyuan*'s first public lecture, which was on electricity, was held in front of an audience of fifty to sixty people.

However, although the library had a collection of several hundred volumes of books on science and technology in Chinese and foreign languages, and the exhibition room held scientific instruments, the *Gezhi shuyuan* did not attract much attention from the locals. The number of visitors and readers was small. When Xu Shou, who managed the reading room, died in 1884, the *Gezhi shuyuan* encountered difficulties, and Fryer seriously considered giving up the enterprise.[56]

When he was pondering the future of the *Gezhi shuyuan,* in 1884, John Fryer established the *Gezhi shushi* (Chinese Scientific Book Depot), which was the outlet for the translations of the Jiangnan Arsenal and other books that Fryer was interested in. The book depot seems to have started to attract attention from those Chinese literati who were interested in books about Western science and technology. According to Fryer's 1887 report, it aimed at "facilitating the spread of all useful knowledge literature in the native language throughout China, and especially of books, maps and other publications of a scientific or technical character." Branches of the book depots were opened in other treaty ports such as Tianjin, Hangzhou, and Shantou in 1886, and then in Beijing, Hankou, Fuzhou, and Xiamen in 1887. Roughly 150,000 volumes of scientific and educational books were sold through the book depot to "the most distant parts of China as well as to Japan and Corea [sic]."[57]

In addition, in 1886, John Fryer and Wang Tao (1928–1897), a reform-minded Chinese literatus who succeeded the deceased Xu Shou in managing the *Gezhi shuyuan*, initiated the "China Prize Essay Contest" (*Gezhi shuyuan keyi*) to attract Chinese literati to write about subjects related to current affairs, especially Western science and technology.[58] In 1888, Fryer wrote in the newspaper *North China Herald* that the contest followed the model of the imperial examinations to assist in missionary efforts to promote Western science, helping to popularize the knowledge contained in translations published between the 1850s and 1880s:[59]

> This scheme [China Prize Essay Contest] is based upon the popular system of writing essays in a high style of composition, in which art the Chinese are known to excel in an extraordinary degree. Every native of any pretensions to scholarship must be able to write and understand an elaborate essay on any given subject. The government examinations, on which all official positions are supposed to depend consist almost wholly of essay writing. Hence the art carried on from century to century, has now reached a degree of perfection with regard to style, arrangement and choice of words which is never thought of nowadays [sic] at any rate in Western lands.

To popularise Western knowledge among the *literati* [original emphasis] it is necessary to take advantage of all such existing national characteristics; and hence it was conceived that in essay writing there existed a most powerful means for inducing the better classes of Chinese to read, think, and write on foreign subjects of practical utility, and thus carry out one of the main objects for which the Polytechnic Institution was founded.[60]

The contest was to be held annually, and Qing government officials would be invited to set essay questions. They would also read and comment on the essays, then offer monetary awards to the prize winners. The involvement of government officials attracted literati who were climbing the ladder of the imperial examinations to submit their essays to be judged. Three major and ten minor prizes would be given, and the names of the winners would be announced in the Chinese and English language newspapers and magazines. The best essays would be published in newspapers such as *Shenbao*. From 1886, the *Gezhi shuyuan* published the collected essays of the top three prize winners in a book.[61] Thus, Fryer was quite satisfied in this endeavor. He reported that "the scheme has succeeded to a degree far beyond the expectations of its originators, and the polytechnic Institution has thus made its influence felt far and wide not only among the students of literature, but among the higher class of officials to whose valuable co-operation and personal interest much of the success is due."[62]

The essay contest was a success. According to Benjamin Elman, during 42 contests held between 1886 and 1893, there were a total of 2236 candidates who submitted essays, and an average 46.1 percent of the contestants were awarded prizes. Fifty-three contestants were frequent participants.[63] Fryer's intention to follow the model of the imperial examinations and borrow the prestige of government officials did help promote the diffusion of popular science knowledge among Chinese literati in the treaty ports. Indeed, these essays give us a chance to explore the extent to which the Chinese literati had absorbed the science and technology translations of the past three decades. Here, we only focus on the essays on heat.

In the 1890 contest, Li Hongzhang set three essay questions, all of which were meant to test contestants' knowledge about recently introduced Western science.[64] One of the questions asked on what basis Westerners measured sound, heat, light, and electricity. On the question about measuring heat, we may assume that Li Hongzhang, who was no physicist, probably meant to ask how Westerners measured temperature.

Sixteen essayists earned the first-class prize, and the essays of the top three were published. Due to limited space, I will discuss the first two winners.

Yang Yuhui, a licentiate (*shengyuan*) from Guangdong who won the first prize, answered the question with a lengthy and elaborate essay. He stated that, according to Westerners, everything in the universe contains heat; heat radiates and diffuses; and the nature of heat is extremely subtle and complex. Because of heat, water can be turned into steam. Westerners argue that to explore the amount of heat an object can absorb, one has to understand the effect of heat. There are several factors that affect the measurement of heat. The first is heat conduction. For example, if one immerses a copper cylinder that is wrapped with wax at the top in a copper basin that is filled with hot water, heat will be transmitted from the hot water to the top of the copper cylinder and melt the wax. Heat also radiates: a hot object, such as an iron casket that is filled with hot water, a red hot iron ball, or a human body, can radiate heat. Even a cold body does the same. Different substances radiate heat differently. (Then Yang gives descriptions of two different instruments that measure radiant heat.) All objects absorb heat, but dark-colored objects absorb more heat than light-colored ones under the sun. Additionally, different substances have different heat capacities (*rongre zhi lü*). To measure them, one can use an instrument that is composed of three layers of iron sheets. The two chambers of empty space are filled with crushed ice. The ice of the outer chamber acts as a heat insulator. One can measure an iron ball's heat capacity by heating it to 100 degrees and then allowing to cool down to 32 degrees in that iron vessel. By weighing the melted ice, one can measure the iron ball's capacity for heat. The essay continues to say that one can use a thermometer to measure the degree of heat (*redu*) by observing the level of the mercury. The essay contains illustrations of an instrument that measures radiant heat, the calorimeter (Fig. 4.2), as well as the mercury thermometer.[65]

Yang's essay, which reprints diagrams of the instruments he mentioned, is lengthy but conceptually rudimentary. Its discussion contains three concepts, namely heat conduction, radiant heat, and specific heat, as well as the method of measuring them. However, he does not present any explanation for why these phenomena have to do with measuring temperature, nor does he mention thermal expansion and contraction, which is the basis of the thermometer. He also fails to recognize that measuring specific heat is different from measuring temperature.

Fig. 4.2 The illustration of the calorimeter provided by the essayist Yang Yuhui in the 1890 Chinese Prize Essay Contest. (Source: *Gezhi shuyuan keyi*, Volume 2 (Shanghai: Shanghai kexue jishu wenxian chubanshe, 2016), p. 373.)

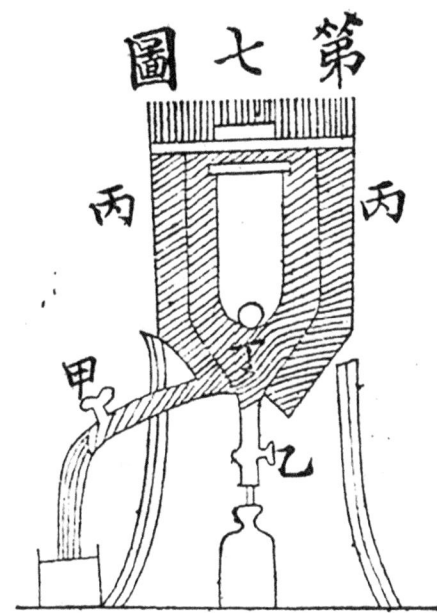

The second prize winner, Wang Fucai, a licentiate from Zhejiang province, submitted a slightly shorter essay that states that China already had a long and subtle tradition in the study of heat, but Westerners seem to have an edge. According to Westerners, the sun is the source of heat, and ice also contains heat. Friction can generate heat; therefore, striking a steel knife against a piece of flint can create fire. Compressing solids, fluids, or air can also generate heat, as can the chemical reaction between sulfuric acid and water. Additionally, different substances transmit heat at different speeds. When a hotter object is in contact with a colder object, heat would flow from the hotter object to the colder one until they each had the same amount of heat. Wang states that the most common instrument for measuring heat is the mercury thermometer. By observing the level of the mercury in the glass tube, one can measure the amount of heat in the air and in all substances. The essay also describes the pyrometer, which contains a platinum wire in an iron tube and can be placed in a furnace to measure temperatures beyond the boiling point of mercury.[66]

Unfortunately, the mention of the source of heat is irrelevant to how to measure temperature. Although thermal conduction has something to do with the thermometer, the description of two different kinds of thermometer does not explain thermal expansion and contraction.

Both essays earned praise from Governor-general Li and Wang Tao for their knowledgeability and talent in essay writing. Neither contestant employed the traditional conceptions of *qi*, *yin-yang*, or the five phases in their arguments. Nevertheless, they did not answer Li Hongzhang's question directly. If one finds their answers uninspiring, they are indeed the authors' uncritical summaries of what they had read. This exposes the shortcomings of following the model of imperial examinations and using the prestige of senior government officials as a means of promoting new scientific knowledge.

The purpose of the imperial examinations was to select government officials through a test of literary competence, moral soundness, and ideological commitment. For centuries, the system had been closely tied to the Chinese monarch and hence, obtaining a degree provided intellectual and social prestige and, ultimately, political power. In the Ming and early Qing dynasties, questions about natural studies and calendrical knowledge were included in the examination questions, but, as Benjamin Elman has documented, the change of politics and intellectual trends reduced the role of natural studies in the examinations.[67]

Although Chinese literati had the opportunity to read the popular science texts published after the 1850s and those in Shanghai might even have attended the *Gezhi shuyuan*'s lectures and visited their exhibitions, their understanding of heat was still limited. The core of the problem was that thermal physics had gone far beyond simply observing thermal phenomena in everyday experiences and extrapolating rough ideas from those observations. Rather, controlled experiments for testing certain thermal phenomena with careful measurements and then formulating equations to provide mathematical descriptions had become the norm among physicists. However, in late nineteenth-century China, except for those who were educated in either the School of Combined Learning or the Dengzhou missionary school, the literati had little chance in receiving a scientific education. Although popular science texts did extend the horizons of Chinese literati's knowledge of nature, reading texts in itself would not be enough for China to absorb the science that underpinned steam engine technology.

One might compare the Essay Contest with prize contests held by academies or learned societies in western Europe. Unfortunately, the difference cannot be starker. We can take three prominent learned societies as examples. The Royal Society in London was formed in 1660 by a group of physicians and natural philosophers, including the experimental chemist Robert Boyle, who had been holding regular meetings to discuss their experimental discoveries. In 1662, the society was granted a royal charter by British King Charles II, who was an earnest patron of science. The French Academy of Sciences (the French Institute after 1795) was established by the French King Louis XIV as a government organ in 1666 to promote science and provide the monarchy with policy advice. Similarly, the Berlin Academy was formed by then-Prince Frederick III of Brandenburg upon the advice of the mathematician and philosopher Gottfried Leibniz to promote the studies of science and the humanities. Although they were organized and functioned differently, each of these societies consisted primarily of scientists.

The Royal Society started to offer honorary prizes to experimentalists in 1731. The Society's council would choose the recipients of the award. The British government also offered an award to those who could solve specific questions, most famously the Board of Longitude, which was established by the British government in 1714 to sponsor the discovery of longitude at sea until 1828. In 1750, the Berlin Academy started to offer medals and prizes to those who answered the specific questions set by the academy's commission. The French Academy started a similar scheme in 1778. The French competition greatly benefited the advance of the science of heat in particular because the 1811 competition's subject was heat conduction, and the 1812 competition was on specific heat. Both had huge impacts on the science of heat.[68] These institutions still offer prizes and medals to those who make significant contributions to science today.

In contrast, the *Gezhi shuyuan* was not a learned society. Its board consisted of foreign merchants in Shanghai, foreign missionaries, and local Chinese dignitaries, and it enjoyed patronage from Qing government officials. Ultimately, Fryer and Wang Tao were promoting knowledge about foreign countries. Popular science was just one topic. Hence, essay questions ranged from China's international trade policy and comparisons between Western and Chinese learning to the question of whether Western language was better than Chinese when studying Western learning.[69] Furthermore, those government officials who were invited to set essay questions were themselves no better informed in science than the contestants. By following the model of the imperial examinations, the essay contest was more a competition of literary competency than scientific reasoning and discovery. Although such a competition could have helped to promote popular science, which in itself might have encouraged the study of real science, the radical education reform that would be necessary for science to take root in Chinese society did not occur until the early twentieth century. In reality, none of the contestants left a significant contribution to the development of science and technology in China.

ENERGY, FORCE, OR WORK?

In 1894, the Jiangnan Arsenal published the *Binchuan qiji* (*Warship Steam Engine*), which was translated by John Fryer and Hua Beiyu, a Chinese literatus about whom I know nothing.[70] The source text was the second edition of the British engineer Richard Sennett's *Marine Steam Engine*, which was first published in London in 1882. Like Main and Brown's *The Marine Steam Engine*, it was meant to provide general knowledge about engine-room management to British Royal Naval College students and Royal Navy officers. It was revised and republished ten times until the 1910s. Its popularity likely had something to do with how effectively it drew from the scientific research on energy.

Sennett's *Marine Steam Engine* has six parts, and each part has four chapters. Part 1 is an introduction to the concepts of work, efficiency, and energy, as well as the nature of heat and its application. It defines "efficiency" as useful work done divided by total energy expended. It also discusses temperature, latent heat, and Joule's theory of the mechanical equivalence of heat and the science of thermodynamics. Part 2 discusses the boiler, focusing on fuel economy, boiler arrangement, and the problem of corrosion and preservation. Part 3 explores the properties of steam and engine efficiency. By quoting Rankine, it focuses on how to increase the efficiency of steam, and hence, the pressure-volume diagram features prominently in the analyses. Part 4 is a detailed discussion of the mechanisms of the valves, gears, cylinders, and condensers. The book also deals with ship hull strength and the propeller mechanism. Hence, Part 5 discusses the scientific principles of a ship hull's co-efficiency in sustaining pressure. Part 6 investigates the question of the indicator diagram, which all engineers had to observe, and other onboard auxiliary machinery. Notably, mathematics lies in the heart of calculating work and efficiency in issues related to the boiler, the use of steam in the cylinder, and the ship's hull.⁷¹ As such, the book contains algebraic, logarithmic, and trigonometric equations for calculation.

The contents of Sennett's textbook were far more updated than Main and Brown's work. Fryer and Hua Beiyu's translation reproduces all the illustrations of the *Marine Steam Engine*. Moreover, they translated the equations as Wylie and Xu Shou did. It is not a literal translation but still closely follows Sennett's argument. Certain terms deserve our attention. Apart from the words *"yinre"* ("hidden heat") and "latent *xianre*" ("sensible heat"), the term *"relixue"* ("thermodynamics") was invented. This term is still being used today. However, the translation of "efficiency" with the term *"gongyi"* (literally "effectiveness of work") only roughly conveys the idea. Furthermore, to translate "energy," Fryer and Hua Beiyu used the term *"li,"* which literally means "force." Because the book was meant to be an engine room manual, this mistranslation would have been a serious error.

Although Sennett's *Marine Steam Engine* gives clear definitions for "force," "work," "foot-pound"—the unit of work—and "efficiency," it does not include any specific passage or sentence to define "energy," as if this was a commonly-known concept. In contrast, William Rankine's *A Manual of the Steam Engine* (1859) clearly defines energy as "the capacity to perform work," and then provides a calculus equation to denote it.⁷²

Sennett's passages that define the force, work, and efficiency of the steam engine read:

> Force: Force is that which acts in producing or resisting motion in a body… and [may be] expressed generally as being equivalent to so many pounds weight.

> Work: A force is said to perform work when by its action resistance is overcome and motion produced. This union of pressure and motion is essential to the conception of work. … Work is measured and expressed as a definite quantity by multiplying the amount of the resistance overcome—or, in other words, the force opposing the motion—by the distance through which the force acts.

> Efficiency: In every machine there are always certain causes acting that produce waste of work, so that the whole work done by the machine is not usefully employed, some of it being exerted in overcoming the friction of the mechanism and some wasted in various other ways. The fraction representing the ratio that the useful work done bears to the total energy

expended by the machine is called the efficiency of the machine; or Efficiency = Useful work done / Total energy expended.[73]

Nevertheless, Fryer and Hua's translation is quite confusing:

> *Li* (force): The meaning of force is that it makes a body move or prevents it from moving.
>
> *Gong* (work): Work is performed by overcoming resistance. There must be two events: obvious force [action] and obvious motion. ... The method to measure work is to multiply the amount of force that overcomes resistance or the force opposing motion by the distance through which the force acts.
>
> *Gongyi* (efficiency): There must be wasted *li* [work] in every machine, therefore the work [*gong*] done is not equal to the *li* used, because machines unavoidably [suffer from] friction and other issues, and [hence] causes a certain amount of *li* to be unable to achieve *gongyi* [efficiency]. If [one] divides the work done by the *li* [energy] used, and then [one] gets a fraction of zero. That is the machine's *gongyi* [efficiency]. Or [one can] calculate the ratio of *gongyi* by dividing the work done by the *li* used.[74]

Although both force and work are clearly defined in the first two passages, the word "*li*" in the third passage actually denotes two different concepts: work and energy. Chinese readers would have understood "*li*" as "force." If they tried to calculate efficiency through that description, they would certainly be confused by which value should be applied to the word "*li*". Using the wrong value creates completely different results. One cannot help but suspect that Fryer did not have a good grasp of the word "energy," hence, his communication with Hua Beiyu created this serious confusion. Indeed, a late Qing literatus would earn praise for his knowledgeability by talking about the *li* discussed in the *Science Primers*, but a technician who claimed to have obtained his knowledge from the *Bingchuan qiji* should only be allowed to manage a model steamboat like Ding Gongcheng's.

Fortunately, no qualified Chinese was likely to claim to build his technical skills on the *Bingchuan qiji*. This had little to do with the usefulness of book's content but rather was a result of how China's naval forces were formed. China's two main naval fleets, the *Beiyang* (Northern Sea) and the *Nanyang* (Southern Sea) fleets, were staffed mostly by Chinese officers trained at the Fuzhou Navy Yard's School of Navigation, which taught students in English without resorting to any translated textbooks in

Chinese. (The history of the Fuzhou Navy Yard's school will be discussed in the next chapter). Foreign naval advisors were recruited to train Chinese officers, and hence, as in the case of the *Qiji faren* mentioned above, no one would have used the *Bingchuan qiji* as teaching material.

In 1894, the Sino-Japanese War broke out over influence in Korea. China's best-equipped *Beiyan* Fleet and the Huai Army were defeated by the Japanese Imperial Navy and Army. As Benjamin Elman has argued, the humiliation turned the mood among Qing government officials and Chinese literati against efforts to introduce Western science and technology into China.[75] Indeed, late Qing literati and government officials had debated whether to put the subjects of natural studies and mathematics into the examinations.[76] Before politics moved ahead, textbooks were still being translated by the Jiangnan Arsenal and missionary presses. Radical changes in education did not happen until the early twentieth century.

Luan Xueqian, the Chinese editor who assisted John Fryer in editing the *Gezhi huibian*, began to teach chemistry after 1895 when the *Gezhi shuyuan* started to provide science courses to the public. In 1924, he outlined the problem of the literati culture of reading science texts:

> In recent years, many books on chemistry had been translated in China, but it was regrettable that those who had browsed through them had obtained only superficial knowledge, and were not able to gain a full understanding [theoretically]. There were amongst such people one or two brilliant scholars who were able to understand things, but, as they had never stooped to try out experiments themselves, in the end this created an obstacle [for them]. At that time, there were many who were resolute in pursuit of this subject.[77]

In other words, Chinese literati might have, by following the centuries-old tradition of *gewu* (investigating things) and *qiongli* (exhausting principles), observed and thought about natural phenomena, they still had only limited knowledge about modern science, which would not have enabled them to play a significant role in importing steam engine technology.

Conclusion

In the 1860s, thanks to the Jiangnan Arsenal's translation project, three marine steam engine textbooks were published for the Chinese readership. Those textbooks, such as the *Qiji faren*, provided succinct knowledge about caloric, latent heat, and heat capacity to their readers, but mainly dealt with technical knowledge about managing the marine steam engine. Calculating and controlling temperature and pressure was those textbooks' major concern, and mathematics were required if readers wanted to gain a good grasp of those issues. Unfortunately, these textbooks were outdated by the 1860s standard and were not used in any schools, so their impact on the Chinese literati's understanding of heat and work could not be gauged.

More popular science texts were published in the journal *Gezhi huibian*, which was founded in 1877. The magazine published essays on the more up-to-date dynamical theory of heat, and the simple expositions of the thermal phenomena could have helped their readers to understand thermal phenomena more than the Jiangnan Arsenal translations did. The magazine attracted Chinese readers in the treaty ports, and every issue sold at least a few thousand copies. The Jiangnan Arsenal and the Imperial Chinese Maritime Customs also published popular science texts by translating the *Science Primers* series. The volume on physics offered a simple explanation for the concept of energy but unfortunately had difficulty in creating a new Chinese term for it, causing confusion between the concepts of force and energy. The same problem occurred in the Jiangnan Arsenal's translation of another marine steam engine textbook, the *Bingchuan qiji*.

The achievements of late Qing popular science translations can be found in the essays submitted to the China Prize Essay Contest, which was organized by John Fryer and Wang Tao. The essayists, who had learned from popular science texts, no longer used the traditional *qi, yin-yang*, or the five phases in explaining thermal phenomena. They knew about heat conduction and radiant heat; they also knew that the thermometer and the calorimeter could be used in measuring temperature and specific heat. This sounds rather basic from our perspective, but it was an important change for the vast majority of late Qing Chinese literati who did not have a chance to attend new schools.

Unfortunately, the Chinese literati's understanding of thermal phenomena was still limited, partly because those translations were done in an unorganized manner. Translators selected source texts either under government officials' demand or due to their own desire to promote popular science. They could not become a coherent body of scientific knowledge either purely to develop thermal physics or to transfer steam engine technology. Furthermore, because science was rapidly changing, late Qing efforts to translate the science of the steam engine faced an impossible task in catching up, especially when none of the translators had any training in science or engineering, the use of their translations was very limited.

The fundamental problem was that Chinese literati could only learn popular science by reading texts. Modern science would have been better transmitted in an institutional context in which schools were established from the basic to the advanced levels and students were guided by qualified teachers through experimentation. Unfortunately, China's educational reform in the second half of the nineteenth century was limited to a few new government and missionary schools. As such, China's domestically trained engineers who were capable of designing and building steamers could only be found in the Naval School at the Fuzhou Navy Yard, which completely broke away from the Chinese tradition of natural studies. That is the topic of the next chapter.

Notes

1. Xiong Yuezhi, *Xixue dongjian yu wan Qing shehui*.
2. For Xu Shou and Hua Hengfang's careers in Western Science, see David Wright, "Careers in Western Science in Nineteenth-Century China: Xu Shou and Xu Jianyin," pp. 49–90; Yang Gen (ed), *Xu Shou he Zhongguo jindai huaxue shi* (Beijing: Kexue jishu chubanshe, 1986).
3. Cheng Peifang, "Xu Xuecun xiansheng xu," *Gezhi huibian*, October 1877, reproduced in Wang Guangren (ed), *Zhongguo jindai kexue xianqu: Xushou fuzi yanjiu* (Beijing: Qinghua daxue chubanshe, 1998), p. 121.
4. Xu Shou's letters are reprinted in Jiang Shuyuan, "Xu Shou de liangfeng qinbixin" (Two letters from Xu Shou written in his own hand), in *Zhongguo keji shiliao* 5:4 (1984), pp. 52–54. The English translation is based on David Wright, "Career in Western Science in Nineteenth-Century China," p. 58.
5. David Wright, "Career in Western Science in Nineteenth-Century China," p. 58.
6. Unfortunately, due to a lack of documents, we cannot further explore the steamship building technology at the Jiangnan Arsenal. The main source

for the history of the Jiangnan Arsenal is the *Jiangnan zhizao ju ji (Records of the Jiangnan Arsenal)*, which was compiled by the general manager of the Arsenal in 1905. Historians have summarized the Arsenal's shipbuilding projects based on the *Records*; for example, Meng Yue, "Hybrid Science versus Modernity: The Practice of the Jiangnan Arsenal, 1864–1897," *East Asian Science, Technology, and Medicine* 16:9 (1999), pp. 13–52. For further discussion of the firearm production at the Arsenal, see Thomas L. Kennedy, *The Arms of Kiangnan: Modernization in the Chinese Ordnance Industry 1860–1895* (Boulder, Colorado: Westview Press, 1978).
7. *Yangwu yundong* vol. 4, p. 79.
8. For more details, see Xiong Yuezhi, *Xixue dongjian yu wan Qing shehui*, pp. 392–437.
9. John Fryer, 'An Account...', *North China Herald*, 29th Jan 1880.
10. For more details about John Fryer, see Adrian A. Bennet, *John Fryer: The Introduction of Western Science and Technology into Nineteenth-Century China* (Cambridge, Mass: East Asian Research Center, Harvard University, Distributed by Harvard University Press, 1967); David Wright, *Translating Science*, pp. 100–148; Xiong Yuezhi, *Xixue dongjian yu wan Qing shehui*, pp. 450–465.
11. Wang Yangzhong, '*Liuhe congtan* zhong de jindai kexue zhishi ji qi zai Qing mo de yingxiang', *Zhongguo keji shiliao* 20:3 (1999), pp. 211–226.
12. Wylie had worked with Wang Tao, a reform-minded Chinese literatus, on *The Origins of Western Astronomy* (*Xiguo tianxue yuanliu*) and *An Elementary Introduction to Mechanics* (*Zhongxue qianshuo*, from *Chamber's Information for the People*) (1856). He also worked with Li Shanlan on *On Astronomy* (*Tantian*, from John Hershel's *Outline of Astronomy*) (1859), the *Supplement to Euclid's Elements* (*Xu jihe yuanben*) (1856), *Algebra* (*Daishuxue*, from Augustus De Morgan's *The Elements of Algebra*) (1859), and *Analytical Geometry and Calculus* (*Dai weiji shiji*, from Elias Loomis' *Elements of Analytical Geometry and of the Differential and Integral*) (1859). All were published by the London Missionary Society Press in Shanghai. For more details, see Han Qi, "Chuanjaishi Weilie Yali zai Hua de kexue huodong," *Ziran bianzhengfa tongxun* 20:4 (1998), pp. 57–70; Wang Xiaoqin, 'Weilie yali de xueshu shengya, *Zhongguo keji shiliao* 20:1 (1999), pp. 17–34; Xiong Yuezhi, *Xixue dongjian yu wan Qing shehui*, pp. 147–151.
13. John Fryer, "Scientific Terminology: Present Discrepancies and means of Securing Uniformity," in *Records of the General Conference of the Protestant Missionaries of China Held at Shanghai, May 7–20, 1890* (Shanghai: American Presbyterian Mission Press, 1890), p. 536.
14. David Wright, *Translating Science*, pp. 327–365.

15. The *Huaxue jianyuan* is probably one of the most-studied late Qing translations of a Western scientific text because of its influence in Chinese chemical nomenclature. Its source text was the 1858 edition of a popular applied chemistry textbook, *Wells's Principles and Applications of Chemistry* by the American natural philosophy professor David Wells. Fryer and Xu actually only translated the inorganic section of the book and left out the parts on the theoretical principles of chemistry and organic chemistry. For more details, see Zhang Zigao and Yang Gen, "Cong *Huaxue chujie* he *Huaxue jianyuan* kan Zhongguo zaoqi fanyi de huaxue shuji he huaxue minci," in Yang Gen (ed.), *Xu Shou he Zhongguo jindai huaxue shi*, pp. 105–118; David Wright, "The Great Desideratum: Chinese Chemical Nomenclature and the Transmission of Western Chemical Concepts," *Chinese Science* 14 (1997), pp. 35–70; Zhang Hao, "Fu Lanya di huaxue fanyi yuanze he linian," *Zhongguo keji shiliao* 21:4 (2000), pp. 297–306; Zhang Hao, "Zai chuantong yu chuangxin zhijian: shijiu shiji de zhongwen huaxue yuansu minci," *Huaxue jiaoyu* 59:1 (2001), pp. 51–59; Wang Yangzong, "A New Inquiry into the Translation of Chemical Terms by John Fryer and Xu Shou," in Michael Lackner, Iwo Amelung and Joachim Kurtz (eds.) *New Terms for New Idea: Western Knowledge and Lexical Change in Late Imperial China* (Leiden: Brill, 2001), pp. 271–283.
16. Xiong Yuezhi, *Xixue dongjian yu wan Qing shehui*, p. 429, 431.
17. Thomas J. Main and Thomas Brown, *The Marine Steam-Engine: Designed Chiefly For The Use of The Officers of Her Majesty's Navy* (London: Longman, Brown, Green and Longmans, 1855); Sun Lei and Lu Lingfeng, "Jiangnan zhizao ju zhengqiji yizhu diben kao," *Wakumon* 20 (2011), p. 39.
18. Feng Shanshan, "Jindai xifang rexue Zai Zhongguo de chuanbo, 1855–1902," p. 137.
19. A list of new terms introduced by the *Qiji faren* can be found in Sun Lei, "Jiangnan zhizaoju zhengqiji yizhu yanjiu," master thesis, University of Science and Technology of China, (2011), pp. 31–36.
20. Mei Yina (Thomas J. Main) and Bai Laona (Thomas Brown), *Qiji faren*, translated by Weilie Yali (Alexander Wylie) and Xu Shou, the 1871 edition kept in library of the Institute for the History of Natural Science, Chinese Academy of Natural Sciences.
21. The Jiangnan Arsenal's own translations of mathematical texts came later. Fryer and Hua Henfang collaborated on the translation of four mathematical treatises: *Daishuxue* (1874), the translation of William Wallace's *Algebra*; *Weiji shuoyuan* (1874) from William Wallace's *Fluxions*, which was an entry from the eighth edition of the *Encyclopaedia Britannica*; *Sanjiao shuli* (1878) from John Hymers's *A Treatise on Plane and Spherical Trigonometry*; and *Daishu nanti* (1878) from *A Companion on Wood's Algebra*.

22. John Bourne, *A Catechism of the Steam Engine: A Handbook of the Steam Engine* (London: Longman, Green, Longman, Roberts, & Green, 1865); Nicholas Procter Burgh, *Pocket Book of Practical Rules for the Proportions of Modern Engines and Boilers for Land and Marine Purpose* (London: E.&F.N. Spon, 1864).
23. Lynwood Bryant, "The Role of Thermodynamics in the Evolution of Heat Engines" *Technology and Culture* 14:2 Part 1 (1973), pp. 152–165.
24. For more details, see Knight Biggerstaff, *The Earliest Modern Government Schools in China*.
25. Xiong Yuezhi, *Xixue dongjian*, pp. 266–277; David Wright, *Translating Science*, pp. 308–311.
26. Fu Lanya (John Fryer), "Jiangnan zhizao zongju fanyi xishu shilue", *Gezhi huibian*, 6th Lunar Month, 1880.
27. Huang Qingcheng, "Zhongxi putong shumu biao," quoted in Xiong Yuezhi, *Xixue dongjian yu wan Qing shehui*, p. 406. Huang Qingcheng (1863–1904), a traditional college teacher in Shanghai in the 1880s, was interested in Western learning and mathematics. For a short introduction to the life and work of Huang Qingcheng, see Kang Xiaoyu, Songyiwen, Yao Yuan, "Wan Qing sanzhong *Suanxuebao* yu shuxue zhuanye qikan dansheng de yiyi," *Xibei daxue xuebao* (Natural Science Edition), 47:1 (2017), pp. 146–151.
28. For a list of the Jiangnan Arsenal's publications, see Xiong Yuezhi, *Xixue dongjian*, pp. 423–432.
29. The *Tongwen guan* in Beijing had its own publication projects, primarily publishing books on elementary chemistry. Anatole Billequin (1837–1894), whose Chinese name was Bi Ligan, was the professor of chemistry at the *Tongwen guan*. For teaching, he wrote the *Huaxue zhinan* (*A Guide to Chemistry*) on the basis of an elementary level chemistry textbook, *Leçons élémentaires de chimie* (1853) by the French chemist Faustino Malaguti. In 1888, Billequin published a more advanced textbook *Huaxue chanyuan* (*An Explanation of Chemistry*) based on books by the German chemist Carl Remigius Fresenius. For more details, see Zhang Hao, "Bi Ligan (Anatole Billequin): Tongwen guan diyiwei huaxue jiaoxi yu jinxiandai Zhongguo huaxue," *Zhonghua kejishi xuehui huikan* 8 (2005), pp. 33–42.
30. The missionary school Luan Xueqian attended was *Wenhui guan* (Literary Association School), which was established as a boy's boarding school by the American missionary Calvin Wilson Mateer (1836–1908) in 1864. In 1873, it became Tengchow Boy's High School, and in 1882, it became Shantung College. For more details, see David Wright, *Translating Science*, pp. 320–325.
31. "Qiji yaoshuo," *Gezhi huibian*, 2nd Lunar Month 1876.
32. "Gezhi luelun xu di wu Juan," *Gezhi huibian*, 6th Lunar Month, 1876.

33. Xiong Yuezhi, *Xixue dongjian*, p. 339.
34. David Wright, *Translating Science*, pp. 163–173; Xiong Yuezhi, *Xixue dongjian*, pp. 334–338.
35. It was also translated by the Jiangnan Arsenal as the *Gezhi xiaoyin*. Thomas Henry Huxley, *Science Primers: Introductory* (London: Macmillan & Co., 1880).
36. Crosbie Smith, *The Science of Energy: A Cultural History of Energy Physics in Victorian Britain*, pp.253–255, 280.
37. Balfour Stewart, *Science Primers: Physics* (London: Macmillan and Co., 1872), p. i.
38. It was translated again in 1903 by a Chinese man named Fan Diji, about whom I know nothing. For further discussion, see Iwo Amelung, "Some notes on Translations of the *Physics Primer* and Physical Terminology in Late Imperial China," pp. 11–33.
39. Wang Yangzhong, "Wan Qing kexue yizhu zakao," *Zhongguo keji shiliao* 15:4 (1994), pp. 32–40.
40. Wu Xia, "Lun Yingguo Lundunhui chuanjiaoshi Ai Yuese yu Zhong Xi keji jiaoliu," *Harebin xueyuan xuebao*, 31:11 (2010), pp. 121–126.
41. Adrian A. Bennett, *Missionary Journalist in China: Young J. Allen and His Magazines, 1860–1883* (Atlanta: University of Georgia Press, 1983).
42. Balfour Stewart, *Science Primers: Physics*, p. 51.
43. Balfour Stewart, *Science Primers: Physics*, pp. 61–82.
44. Balfour Stewart, *Science Primers: Physics*, pp. 103–104.
45. Wang Yangzong has compared the translations of Huxley's *Science Primers: Introduction* and has found that the Jiangnan Arsenal's translation drops many sentences but keeps the essence of the argument. Edkins' translation is more wordy and less fluent. He sometimes added extra sentences to explain theoretical ideas such as the atom. Both translations are generally faithful to Huxley's text, but some sentences may not accurately transmit the exact meaning of the source text. Wang Yangzong, "He Xuli *Kexue daolun* de liangge Zhong yiben: Jian tan Qing mo kexue yizhu de zhunquexing," *Zhongguo keji shiliao* 21:3 (2000), pp. 207–221.
46. Compare Iwo Amelung's reading of Edkins' prose. Iwo Amelong, "Some Notes on Translations of the *Physics Primer* and Physical Terminology in Late Imperial China," p. 22.
47. Lin Lezhi (Young John Allen) and Zeng Changyan, *Gewu qimeng*, (the 1880 edition kept in the Kuo T'ing-yee Library, Academia Sinica; Ai Yuese (Joseph Edkins), *Gezhi zhixue qimeng*, (the 1898 edition stored in the Kuo Ting-yee Library, Academia Sinica).
48. Xiong Yuezhi, *Xixue dongjian*, p. 376; David Wright, *Translating Science*, pp. 287–288.

49. Pressure would change the boiling point of a fluid. It also discusses heat as the cause of wind, rain, and cloud. Fu Lanya (John Fryer), *Rexue Tushuo* (The 1890 edition kept in the Shanghai Library).
50. The instrument was invented by Antoine Lavoisier and Pierre-Simon Laplace. The basic idea was to measure heat by temperature change. However, it suffered from one problem: the melted water tends be trapped between the pieces of ice.
51. Iwo Amelung, "Naming Physics: The Strife to Delineate a Field of Modern Science in Late Imperial China," in Michael Lackner and Natascha Vittinghoff (eds.), *Mapping Meanings: Translating Western Knowledge into Late Imperial China* (Leiden: Brill, 2004), pp. 381–422.
52. Xu Shou, "Qiji minming shuo," *Gezhi huibian*, 6th Lunar Month, 1876.
53. The proposal was first published in the Chinese newspaper *Shenbao* on 25th March 1873 and later in the English newspaper *North China Daily News*, 5th March 1874. For more details about the Polytechnic, see Xiong Yuezhi, *Xixue dongjian yu wan Qing shehui* (Beijing: Zhongguo renmin daxue chubanshe, 2010), pp. 278–307; David Wright, "John Fryer and the Shanghai Polytechnic: Making space for Science in Nineteenth-Century China", *The British Journal for the History of Science* 29:1 (1996), pp. 1–16; Knight Biggerstaff, "Shanghai Polytechnic Institution and Reading Room: An Attempt to Introduce Western Science and Technology to the Chinese," *Pacific Historical Review* 25:2 (1956), pp. 127–149.
54. David Wright, "John Fryer and Shanghai Polytechnic," p. 8.
55. A prominent Chinese merchant and comprador Tang Tingshu, better known as Tong Kingsing, was also on the board. Both Medhurst and Tang were major fund raisers for the *Gezhi shuyuan*. Xiong Yuezhi, *Xixue dongjian*, p. 280.
56. David Wright, "John Fryer and the Shanghai Polytechnic," pp. 10–12.
57. John Fryer, "Report of the Chinese Scientific Book Depot, Shanghai, 1887," *North China Herald*, 28th December 1887.
58. Wang Tao, whose career was closely linked with promoting Western science and technology from the 1840s, had traveled to France and Britain between 1867 and 1870. Wang lived in Hong Kong from 1870 to 1884, editing a Chinese newspaper in which he published many articles promoting his idea of political reforms and earned attention from Li Hongzhang, the governor-general of Zhili and the most influential senior official in the self-strengthening politics, as well as Meiji reformers in Japan. Hence, he was invited to visit Japan in 1879. Wang returned to Shanghai in 1884, when he was invited by the board of the Polytechnic to be the manager of the institution. For more information on Wang Tao, see Paul A. Cohen, *Between Tradition and Modernity: Wang T'ao and Reform in Late Ch'ing China* (Cambridge, Mass.: Harvard University Press, 1988).

59. For an overview of the history of the essay contests, see David Wright, "John Fryer and the Shanghai Polytechnic,", pp. 1–16; Benjamin A. Elman, "The China Prize Essay Contest and the Late Qing Promotion of Modern Science," in Ho Yi Kai (ed), *Science in China, 1600–1900: Essays by Benjamin A. Elman* (Hackensack, NJ, World Century and World Scientific, 2015), pp. 139–171.
60. John Fryer, "Chinese Prize Essays: Report of the Chinese Prize Essay Scheme in connection with the Chinese Polytechnic Institution and Reading rooms, Shanghai, for 1887 and 1886," *North China Herald*, 25th January 1888.
61. The essays were collected by Wang Tao and published in Shanghai in 1897 as *Gezhi keyi huibian* (*Collected Essays of the China Prize Essay Contest*).
62. John Fryer, "Chinese Prize Essays: Report of the Chinese Prize Essay Scheme in connection with the Chinese Polytechnic Institution and Reading rooms, Shanghai, for 1887 and 1886," *North China Herald*, 25th January 1888.
63. Benjamin A. Elman, "The China Prize Essay Contest and the late Qing Promotion of Modern Science".
64. Li's first question was about whether all of the sixty-four elements in chemistry could be found in China and how one might explain the "essence (*ti*) and use (*yong*)" between Chinese and Western learnings. The third question required contestants to discuss the positive (*yang*) and negative (*yin*) metals used in a battery. *Gezhi shuyuan keyi* (Shanghai: Shanghai kexue jishu wenxian chubanshe, 2016) vol. 2, p. 327.
65. *Gezhi shuyuan keyi*, vol. 2, pp. 356–357.
66. *Gezhi shuyuan keyi*, vol. 2, pp. 400–403.
67. Benjamin A. Elman, *A Cultural History of Civil Examinations in Late Imperial China* (Berkeley: University of California Press, 2000).
68. The 1811 competition was won by Joseph Fourier, whose research was later published as *Théorie analytique de la chaleur* (*The Analytical Theory of Heat*). The 1812 competition was won by François Delaroche and Jacque-Étienne Bérard. Their paper was published as "Mémoire sur la détermination de la chaleur spécifique des différens gaz." *Annales de chimie et de physique*, 85 (1813), pp. 72–110, 113–182.
69. In the 1889 contest, Li Hongzhang asked contestants to compare the history of Chinese and Western learnings. In the same year, John Fryer questioned the benefits and drawbacks of using Chinese or Western language to study Western learning. *Gezhi shuyuan keyi*, vol. 2, pp. 3–4.
70. We do not know if Hua Byiyu was related to the mathematician Hua Hengfang, whose brother Hua Shifang was also a mathematician.
71. William Rankine's formulae are prominent in the book's calculations.

72. William John Macquorn Rankine, *A Manual of the Steam Engine and Other Prime Movers* (London: Charles Griffin and Company, 1859), p. xxxii.
73. Richard Sennett, *The Marine Steam Engine: A Treatise for the Use of Engineering Students and Officers of the Royal Navy* (London: Longmans, Green, and Co., 1885), p. 27
74. *Bingchuan qiji*,1:18a-9b (the 1894 edition kept in the Kuo Ting-yee Library, Academia Sinica).
75. Benjamin A. Elman, "Naval Warfare and the Refraction of China's Self-Strengthening Reforms into Scientific and Technological Failure, 1865–1895," *Modern Asian Studies* 38:2 (2004), pp. 283–326.
76. For discussions on reforming the imperial examinations, see Benjamin A. Elman, *A Cultural History of Civil Examinations in Late Imperial China*, pp. 585–622.
77. Luan Xueqian, "Gezhi shuyuan jiaoyan huaxue," (1924) quoted in David Wright, *Translating Science*, p. 141.

References

Ai Yuese 艾約瑟 (Joseph Edkins), *Gezhi zhixue qimeng* 格致質學啟蒙 (the 1898 edition kept in the Kyo Ting-yi Library, Academia Sinica).
Amelung, Iwo, "Naming Physics: The Strife to Delineate a Field of Modern Science in Late Imperial China," in Michael Lackner and Natascha Vittinghoff (eds.), *Mapping Meanings: Translating Western Knowledge into Late Imperial China* (Leiden: Brill, 2004a), pp. 381-422.
Amelung, Iwo, "Some Notes on Translations of Physics Primer and Physical Terminology in Later Imperial China," *Wakumon* 或問 11:8 (2004b), pp. 11–34.
Bennet, Adrian A., *John Fryer: The Introduction of Western Science and Technology into Nineteenth-Century China* (Cambridge, Mass: East Asian Research Center, Harvard University, Distributed by Harvard University Press, 1967).
Bennet, Adrian A., *Missionary Journalist in China: Young J. Allen and His Magazines, 1860–1883* (Atlanta: University of Georgia Press, 1983).
Biggerstaff, Knight, "Shanghai Polytechnic Institution and Reading Room: An Attempt to Introduce Western Science and Technology to the Chinese," *Pacific Historical Review* 25:2 (1956), pp. 127–149.
Biggerstaff, Knight, *The Earliest Modern Government Schools in China* (Ithaca, N.Y.: Cornell University Press, 1961).
Bourne, John, *A Catechism of the Steam Engine: A Handbook of the Steam Engine* (London: Longman, Green, Longman, Roberts, & Green, 1865).
Bryant, Lynwood, "The Role of Thermodynamics in the Evolution of Heat Engines," *Technology and Culture* 14:2 Part 1 (1973), pp. 152–165.

Burgh, Nicholas Procter, *Pocket Book of Practical Rules for the Proportions of Modern Engines and Boilers for Land and Marine Purpose* (London: E.&F.N. Spon, 1864).

Cohen, Paul A., *Between Tradition and Modernity: Wang T'ao and Reform in Late Ch'ing China* (Cambridge, Mass.: Harvard University Press, 1988).

Elman, Benjamin A., *A Cultural History of Civil Examinations in Late Imperial China* (Berkeley: University of California Press, 2000).

Elman, Benjamin A., "Naval Warfare and the Refraction of China's Self-Strengthening Reforms into Scientific and Technological Failure, 1865–1895," *Modern Asian Studies* 38:2 (2004), pp. 283–326.

Elman, Benjamin A., "The China Prize Essay Contest and the Late Qing Promotion of Modern Science," in Ho Yi Kai (ed), *Science in China, 1600–1900: Essays by Benjamin A. Elman* (Hackensack, NJ: World Century and World Scientific, 2015), pp. 139–171.

Fryer, John, "Chinese Prize Essays: Report of the Chinese Prize Essay Scheme in connection with the Chinese Polytechnic Institution and Reading rooms, Shanghai, for 1887 and 1886," *North China Herald*, 25th January 1888.

Fryer, John, "Scientific Terminology: Present Discrepancies and means of Securing Uniformity," in *Records of the General Conference of the Protestant Missionaries of China Held at Shanghai, May 7–20, 1890* (Shanghai: American Presbyterian Mission Press, 1890), pp. 531–549.

Fu Lanya 傅蘭雅 (John Fryer), Rexue Tushuo 熱學圖說 (The 1890 edition kept in the Shanghai Library).

Gezhi huibian 格致彙編.

Gezhi shuyuan keyi 格致書院課藝 (Shanghai: Shanghai kexue jishu wenxian chubanshe, 2016).

Han Qi 韓琦, "Chuanjaoshi Weilie Yali zai Hua de kexue huodong 传教士伟烈亚力在华的科学活动," *Ziran bianzhengfa tongxun* 20:4 (1998), pp. 57–70.

Huxley, Thomas Henry, *Science Primers: Introductory* (London: Macmillan & Co., 1880).

Jiang Shuyuan 蔣树源, "Xu Shou de liangfeng qinbixin 徐寿的两封亲笔信" (Two Letters from Xu Shou Written in His Own Hand), in *Zhongguo keji shiliao* 5:4 (1984), pp. 52–54.

Kang Xiaoyu 亢小玉, Songyiwen 宋轶文, Yao Yuan 姚远, "Wan Qing sanzhong *Suanxuebao* yu shuxue zhuanye qikan dansheng de yiyi 晚清三种算学报与数学专业期刊诞生的意义," *Xibei daxue xuebao* 西北大学学报 (Natural Science Edition), 47:1 (2017), pp. 146–151.

Kennedy, Thomas L., *The Arms of Kiangnan: Modernization in the Chinese Ordnance Industry, 1860–1895*, (Boulder, Colorado: Westview Press, 1978).

Lin Lezhi 林樂知 (Young John Allen) and Zeng Changyan 鄭昌棪, *Gezhi qimeng* 格致啟蒙 (the 1880 edition kept in the Kuo Ting-yee Library, Academia Sinica).

Main, Thomas J. and Brown, Thomas, *The Marine Steam-Engine: Designed Chiefly For The Use of The Officers of Her Majesty's Navy* (London: Longman, Brown, Green and Longmans, 1855).

Mei Yina 美以納 (Thomas J. Main) and Bai Laona 白勞納 (Thomas Brown), *Qiji faren* 汽機發軔, translated by Weilie Yali 偉烈亞力 (Alexander Wylie) and Xu Shou 徐壽 (the 1871 edition kept in library of the Institute for the History of Natural Science, Chinese Academy of Sciences, Beijing).

Meng Yue, "Hybrid Science *versus* Modernity: The Practice of the Jiangnan Arsenal, 1864–1897," in *East Asian Science, Technology, and Medicine* 16 (1999), pp. 13–52.

Pu Erna 蒲而捺 (John Bourne), *Qiji biyi* 汽機必以, translated by Fu Lanya 傅蘭雅 (John Fryer) and Xu Jianyin 徐建寅 (the 1872 edition kept in the library of the Institute of Natural Sciences, Chinese Academy of Sciences, Beijing).

Rankine, William John Macquorn, *A Manual of the Steam Engine and Other Prime Movers* (London: Charles Griffin and Company, 1859).

Sennett, Richard, *The Marine Steam Engine: A Treatise for the Use of Engineering Students and Officers of the Royal Navy* (London: Longmans, Green, and Co., 1885).

Shenbao 申報.

Smith, Crosbie, *The Science of Energy: A Cultural History of Energy Physics in Victorian Britain* (Chicago: Chicago University Press, 1998).

Stewart, Balfour, *Science Primers: Physics* (London: Macmillan and Co., 1872).

Sun Lei 孙磊 and Lü Lingfeng 吕凌峰, "Jiangnan zhizao ju zhengqiji yizhu diben kao 江南制造局蒸汽机译著底本考," *Wakumon* 或問 20 (2011), pp. 33–48.

Sun Lei 孙磊, "Jiangnan zhizaoju zhengqiji yizhu yanjiu 江南制造局蒸汽机译著研究," Master Thesis, University of Science and Technology of China, (2011).

The North China Herald

Wang Xiaoqin 汪晓勤, "Weilie yali de xueshu shengya 伟烈亚力的学术生涯," *Zhongguo keji shiliao* 中国科技史料 20:1 (1999), pp. 17–34.

Wang Yangzhong 王扬宗, "*Liuhe congtan* zhong de jindai kexue zhishi ji qi zai Qing mo de yingxiang 六合丛谈中的近代科学知识及其在清末的影响," *Zhongguo keji shiliao* 中国科技史料 20:3 (1999), pp. 211–226.

Wang Yangzhong 王扬宗, "Wan Qing kexue yizhu zakao 晚清科学译著杂考," *Zhongguo keji shiliao* 中国科技史料 15:4 (1994), pp. 32–40.

Wang Yangzhong 王扬宗, "Hexuli *Kexue daolun* de liangge Zhongyi ben: Jiantan Qingmo kexue yizhu de zhunquexing 赫胥黎'科学导论'的两个中译本:兼谈清末科学译著的准确性," *Zhongguo keji shiliao* 中国科技史料 21:3 (2000a), pp. 207–221.

Wang Yangzhong 王扬宗, "He Xuli *Kexue daolun* de liangge Zhong yiben: Jian tan Qing mo kexue yizhu de zhunquexing 赫胥黎科学导论的两个中译本:简谈清末科学译著的准确性," *Zhongguo keji shiliao* 21:3 (2000b), pp. 207–221.

Wang Yangzong, "A New Inquiry into the Translation of Chemical Terms by John Fryer and Xu Shou," in Michael Lackner, Iwo Amelung and Joachim Kurtz (eds.) *New Terms for New Idea: Western Knowledge and Lexical Change in Late Imperial China* (Leiden: Brill, 2001), pp. 271–283.

Wright, David, "Careers in Western Science in Nineteenth-Century China: Xu Shou and Xu Jianjin," *Journal of the Royal Asiatic Society* 5:1 (1995), pp. 49–90.

Wright, David, "John Fryer and the Shanghai Polytechnic: Making space for Science in Nineteenth-Century China", *The British Journal for the History of Science* 29:1 (1996), pp. 1–16.

Wright, David, "The Great Desideratum: Chinese Chemical Nomenclature and the Transmission of Western Chemical Concepts," *Chinese Science* 14 (1997), pp. 35–70.

Wright, David, *Translating Science: The Transformation of Western Chemistry into Late Imperial China, 1840–1900* (Leiden: Brill, 2000).

Wu Xia 吳霞, "Lun Yingguo Lundunhui chuanjiaoshi Ai Yuese yu Zhong Xi keji jiaoliu 论英国伦敦会传教士艾约瑟与中西科技交流," *Harebin xueyuan xuebao* 哈尔滨学院学报, 31:11 (2010), pp. 121–126.

Xiong Yuezhi 熊月之, *Xixue dongjian yu wan Qing shehui* 西学东渐与晚清社会, revised edition (Beijing: Zhongguo renmin daxue chubanshe, 2011).

Xu Shou 徐壽, "Qiji minming shuo 汽機命名說," *Gezhi huibian* 格致彙編, 6th Lunar Month, 1876.

Yang Gen 杨根 (ed.), *Xu Shou he Zhongguo jindai huaxue shi* 徐寿和中国近代化学史 (Beijing: Kexue jishu chubanshe, 1986).

Zhang Hao 張澔, "Fu Lanya di huaxue fanyi yuanze he linian 傅兰雅的化学翻译原则和理念," *Zhongguo keji shiliao* 中国科技史料 21:4 (2000), pp. 297–306.

Zhang Hao 張澔, "Zai chuantong yu chuangxin zhijian: shijiu shiji de zhongwen huaxue yuansu minci 在傳統與創新之間:十九世紀的中文化學元素," *Huaxue jiaoyu* 化學教育 59:1 (2001), pp. 51–59.

Zhang Hao 張澔, "Bi Ligan (Anatole Billequin): Tongwen guan diyiwei huaxue jiaoxi yu jinxiandai Zhongguo huaxue 畢力幹:同文館第一位化學教習與近現代中國化學," *Zhonghua kejishi xuehui huikan* 中國科技史學會會刊 8 (2005), pp. 33–42.

Zhongguo shixuehui 中國史學會 (ed.), *Yangwu yundong* 洋務運動, 8 volumes (Shanghai: Shanghai renmin chubanshe, 1961).

CHAPTER 5

Training Workers and Engineers: The Fuzhou Navy Yard, 1866–1895

Chapter 2 has briefly mentioned the shipbuilding and firearm making of the Jiangnan Arsenal. Yet lack of documents limits our understanding of technical capacity of the arsenal.[1] A more detailed record of the role of shipbuilding in the application of Western steam technology and related skills is the Fuzhou Navy Yard. As historians such as David Pong, Marianne Bastid, and others have shown, the navy yard was one of the productive new undertakings of the self-strengthening program. It imported machinery and hired foreign technicians. It built 15 wooden-hulled gunboats equipped with simple and small engines. The technical progress went as far as building steel-hulled cruisers by 1889 that were propelled by sophisticated and powerful engines.[2]

This chapter focuses on how steamship technology was institutionalized by a comprehensive program that put physics, mathematics, technical drawing, machine tools, engineering knowledge at its heart, transforming those artisans and students who entered the navy yard with little knowledge and skills of steamship technology into modern engineers and workers.

INITIATING THE NAVY YARD PROJECT

During the Taiping Rebellion, the anti-Taiping commander Zuo Zongtang had plenty of experiences in employing Western firearms and steamers. He led his troops fighting alongside the Ever-Triumphant Army, a

French-Chinese mixed force that was similar to the Ever-Victorious Army and also employed Western firearms and gunboats.[3] After the Second Opium War, senior officials were discussing the policy of learning to produce Western guns and steamers for self-strengthening, in March 1863, Zuo Zongtang as governor of Zhejiang suggested to the *Zongli yamen* that China should adopt steamship-building as a long-term defense policy.[4] In the same year, he also repeatedly expressed to friends that China needed to build steamers for maritime defense.[5]

His ambition was known to his two French friends, Prosper Giquel and Paul d'Aiguebelle, both French naval officers who served in the Ever-Triumphant Army. By early 1863, through Giquel and d'Aiguebelle, the French naval commander in China invited Zuo to join a partnership of running a French shipyard in Ningbo. Yet, Zuo rejected the offer after he found out the shipyard could only build ship hulls but not steam engines.[6]

Zuo Zongtang was interested in launching a shipbuilding operation. According to Zuo's account, in 1864, as newly appointed governor-general of Fujian and Zhejiang, he hired a Chinese workman to build a small steamboat and put it to the test on the West Lake at Hangzhou. Zuo reported to the imperial court that the boat "cruised slowly."[7] Unfortunately, we do not have any information about how that boat was built. Zuo consulted Giquel and d'Aigubelle, who suggested to Zuo that the boat could be improved if a steam engine purchased from the West was installed. They also showed Zuo plans of steam vessels. Thereupon, Zuo asked for their assistance in his steamship-building project, forming a plan for building a navy yard in the following two years.[8]

In June 1866, Zuo Zongtang memorialized the imperial court, asking for permission to establish a navy yard in the fishing village of Mawei near Fuzhou, Fujian province. He argued that employing steamers was essential in order to defend the country and improve commerce. Yet, to obtain steamers China would have to establish a navy yard, purchase machinery, hire technicians to build steamers and train young Chinese men in the methods of building them. The funds would be provided from the Fujian maritime customs and the transit dues. The imperial court agreed that such a project was urgently needed for self-strengthening and granted permission.[9]

However, soon after the edict, with the Moslem rebellion in the northwest going on, the imperial court appointed Zuo governor-general of Shaanxi and Gansu. It ordered Zuo's successor in Fuzhou to take over the navy yard project.[10] Despite that, after a careful political maneuver, Shen

Baozhen, former governor of Jiangxi and a close associate of Zuo, was appointed director of the navy yard. Soon after that, in December 1866, Zuo Zongtang presented to the imperial court the contract with Giquel and d'Aiguebelle.[11]

FIVE-YEAR CONTRACT

Zuo Zongtang set the contract term for Giquel and d'Aiguebelle for five years. The contract was to establish a navy yard with potential for long-term development. Zuo's ambition was to "learn to build Western machinery and then to build steamers, so that the Chinese people could teach one another [the skills] and turn it into a permanent gain."[12] This would require Giquel and d'Aiguebelle's knowledge about steamship building. Although they themselves were not professional engineers, Giquil and d'Aiguebelle's background of École Navale in France must have had an impact on their design of the project.[13] France had a strong tradition in combining military and engineering training, which put technical drawings, practical training, and theoretical, especially mathematical, foundation in its heart.[14] Besides, they must have learned from their experience in the anti-Taiping campaign that Chinese traditional tools and iron-smelting technology did not have the capacity to build steamships and their engines. Therefore, they were to import machine tools and iron-smelting equipment, build steamers, set up training program, and recruit a large group of foreign technicians whose professions covered every aspect of modern shipbuilding. For training the workmen who had no knowledge of modern shipbuilding, the project started from basic technology, that is, wooden-hulled and small horsepower single-cylinder reciprocating engines. The contract was now:

- To establish schools, one of naval construction and the other navigation, to teach students English and French, mathematics, technical drawing, navigation, and ship-captainship. The students of the school of navigation should be capable of navigating in the open seas within sight of the coast. The students of the school of naval construction should learn enough of mathematics to construct a ship on a given plan
- To build eleven 150-horsepower steamers of the class and five 80-horsepower gunboats within five years. The engines for the first two ships of the former class and for all steamers of the latter are to

be purchased from abroad. The other nine sets of engines will be made at the metal-working factory of the navy yard.
- To purchase machinery for shipbuilding and iron foundering for steam-engine building.
- To recruit foreign technicians to teach the Chinese officers and workers to build steamships according to a given plan and to make machine-tools.[15]

This contract did not give the navy yard the state-of-the-art shipbuilding technology. 80- and 150-horsepower engines were a long way from the mainstream 600- to 800-horsepower engines that were being built in Western shipyards. Furthermore, when warship building in the West had been moving towards iron construction with iron armor, choosing wooden-hull structure seems to have been a conservative decision.[16] Therefore, the immediate objective of the Fuzhou Navy Yard was a long way from contemporary standards of excellence.

After the contract was signed in October, Zuo Zongtang left for Gansu. Giquel and d'Aiguebelle went to France to purchase machinery and hire technicians, while the construction work of the navy yard started under the supervision of local officials.[17] In January 1868, when the workshops were being built, Shen Baozhen gathered the foreign technicians and the Chinese managers to perform the rituals of putting the keel of the first steamer on the slip and laying the foundation stone of the iron foundry. The rituals marked the inauguration of the navy yard, but the navy yard was only completed in around September 1868.[18]

Qing officials kept the management of the navy yard to themselves, although they entrusted technical matters to foreign technicians. Hence, the navy yard was a mixture of a modern factory and the Qing government bureaucracy. Its name followed Chinese government's custom as *Chuanzhen ju* (Shipbuilding bureau). It was also usually referred to as the Fujian bureau (*Min ju*) in Chinese documents.

THE YARD

The scale of the Fuzhou Navy Yard had not been seen in Chinese history. It was located on the left bank of the River Min, half a mile above the Pagoda Anchorage, twenty-two miles from the mouth of the river and nine miles below Fuzhou city. It had 45 buildings for administrative, educational, and production purposes. It covered 118 acres (47 hectares 77 ares) of land, of which the workshops took up 33 acres (18 hectares 8 ares) and the offices, schools, and the accommodation for officials, foreign technicians, and workers took up 85 acres (4 hectares 30 ares).[19]

The navy yard consisted of four major workshops: the metal-working factory, the boiler shop, the fitting shop, and the foundry. These shops were designed to enable the navy yard to build every part of a steamer, from the engine to the barometers. They were connected by tramways with turn tables for transporting materials and parts between workshops.

The metal-working factory (1 acre, or 4190 square meters. Map 5.1 u and v) included heavy forges and rolling mills. These were equipped with six steam hammers, 16 forge fires for heavy work and six furnaces for reheating. The components of the steam engines, shafts and crank shafts were manufactured here. The rolling mill had six furnaces for reheating and four rollers. The rollers rolled out heavy iron plates, angle iron, and small iron objects. The machinery was driven by a 100-horsepower engine. The boiler shop (0.6 acre, or 2400 square meters, Map 5.1 a) had a 15-horsepower engine, which supplied a blast to the forges and movement to the machines. The boilers purchased from Europe would be assembled here. The fitting shop (0.6 acre, Map 5.1 b) and the setting-up shop assembled the components of engines that were either purchased from abroad or produced in the metal-working factory. The offices of engineers and technical drawing were situated in the setting-up shop. The foundry (0.6 acre, Map 5.1 d) was worked by a 15-horsepower engine. It has three cupolas, enabling the foundry to melt masses of 15 tons. It could turn out an average of 12 to 15 tons of castings per week, and among these, parts of the 150-horsepower engines, such as cylinders, and condensers.

Other smaller workshops included a chronometer shop (0.17 acre or 720 square meters, Map 5.1 h), a small forge (0.5 acre or 2160 square meters, Map 5.1 i), which supplied the innumerable smaller articles used in the construction and equipment of ships, a shop for small fittings and locksmith work (0.12 acre or 510 square meters, Map 5.1 j), a mechanical sawmill (0.25 acre or 1020 square meters, Map 5.1 k), and the model room and the joiner's shop (0.36 acre, or 1440 square meters, Map 5.1 l), which produced the models for making engines. The shipbuilding yards included a molding hall (Map 5.1 e), on which could be traced out in full scale the lines of the ships, a crane, capable of lifting 40 tons, and a patent slip on Labat's system (Map 5.1 q), which could haul ships up to the slip from the sideway. It was 350 feet long and was able to sustain a ship of 4000 tons displacement.[20] It was capable of receiving ships of 100 meters

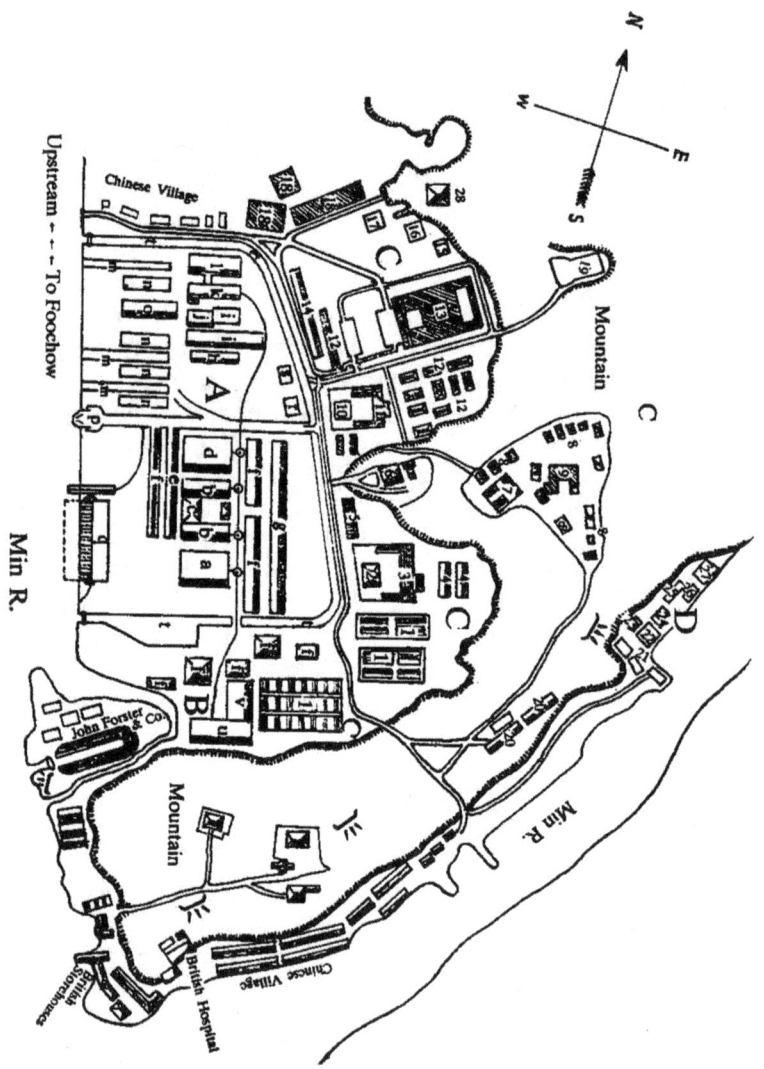

Map 5.1 Plan of the Fuzhou Navy Yard. Source: David Pong, *Shen Pao-chen and China's Modernization in the Nineteenth Century* (Cambridge: Cambridge University Press, 1994), pp. 204–205

5 TRAINING WORKERS AND ENGINEERS: THE FUZHOU NAVY YARD, 1866–1895

A: The Yard

a. Boiler house

b. Fitting shop

c. Setting-up shop

d. Foundry

e. Molding hall

f. Stores

g. Rope and sail shop

h. Shops for chronometers, optical instruments, and compasses

i. Equipment forges

j. Small fitting and locksmith's shops

k. Mechanical sawmill

l. Model room and the joiners' shop

m. Shipbuilding slips

o. Wood-working sheds

p. Mast and small boats workshops

Wharf and caulking machine

q. Labat's patent slip

r. French Directors' Office

s. Office of the Chinese officials

r. Moat

B: Metal-working Forge

u. Forging department

v. Rolling mill

24 Kiln for fire-proof bricks

25 Coking kiln

C: Schools and Living Quarters

1. Chinese workers' accommodation

2. School of Navigation

3. Pupils' accommodation

4. Timber store-houses

5. European technicians' quarters

6. French director's residence

8. Barracks for the guards

9. Residence of the guards' commander

10. The School of Naval Construction

11. Pupils' accommodation

12. Timber store-houses

13. Office of the director and managers

14. European technicians' quarters

15. Theatre

16. Residence of secretary-interpreter and doctor

17. Residence of teachers and secretaries

18. Residence of European technicians

19. The *Tianhou* Shrine

20. Quarters for married workers

28 Second French director's residence

D: Brick and Coking Kilns

21 Overseers' and workers' quarters

22 Moulding shop

23 Brick kiln

26 Shop for preparing fire-proof clay

27 Lime kiln

Map 5.1 (continued)

of keel and of 2500 tons' displacement. Apart from the workshops, buildings for administration and the naval school were also built. There was no dock but there were three building slips (Map 5.1 m).

The scale of the navy yard offered good potential for building much larger vessels than those stipulated by the five-year contract. According to a *North China Herald* report, merchant ships of "330 feet long and of 2,500 tons cargo capacity could be built, engined, and in all respects constructed and completed in the dockyard." It also noted that a warship "of 2,300 tons and 400 H.P. could be built, her engines, boilers and screw manufactured, her armament, sails, cordage, iron mats, and anchors made, and her amidship battery plated with heavy iron laminated plates rolled."[21]

The scale and capacity might not have matched European leading shipbuilders that took decades to develop. For example, in the 1870s, the yard of Barrow Shipbuilding Company, one of the biggest and most modern navy yards in Britain that built both merchant and naval vessels, had 58 acres of land, and was capable of simultaneously building five ships of over 4000 tons together with seven to ten smaller ones.[22] Nevertheless, the Fuzhou Navy Yard was much larger than the 73 acres of the Jiangnan Arsenal and was China's largest modern shipyard.[23]

THE ORGANIZATION

As a government undertaking, the Fuzhou Navy yard had to be staffed by Chinese officials. It would also have to give foreign technicians the power to manage all technical affairs. The imperial court appointed its director, who had the authority to recommend candidates for senior manger to the court for appointment but he could directly appoint junior managers. He had to obtain court permission for launching major project works and his funds were supplied by the Fujian authorities, mainly from the revenues of the Fujian maritime customs and the transit dues.

On the foreign staff side, under the director, were foreign supervisors (*yang jiandu*), Giquel and his deputy d'Aiguebelle, who led all the foreign staff. Directly under them was the chief engineer, who directed and coordinated the works of the workshops.[24] Other foreign technicians supervised the works of all the workshops. Originally Giquel and d'Aigubelle recruited 75 European employees, including an interpreter, teachers, secretaries, foremen, workmen and supernumeraries, but the number was reduced to 52 in 1874.[25] Their specialties included almost all the professions that related to modern shipbuilding: carpenters, boiler makers, molders, blacksmiths, draftsmen, and barometer makers.

On the Chinese side of the staff, there was at least one Chinese supervisor (*tidiao*) who was responsible for administrative work. He would act on behalf of the director when the director was on leave. A chief work supervisor (*zong jian'gong*) might have managed the Chinese work force. It is not clear that he was subordinate to the Chinese supervisors.[26] During the five-year term, the chief work supervisor seems to have acted as a procurement manager, traveling to Southeast Asia to purchase timber. All of these senior managers were imperial official title holders. The supervisor and the chief work supervisor's function would change after the foreign technicians' contract expired in 1874.

Under the Chinese supervisors were junior managers responsible for different administrative tasks. According to a 1886 memorial, the administrative office might have been divided into four main departments: the secretariat (*wen'an chu*) handled the administrative matters of the whole navy yard; the disbursement department (*zhiying chu*) managed accounts, income, and expenditure; the inspecting department (*jicha chu*) inspected the works of the workers and the students; and the work examination department (*kaogong suo*), that was established in 1869 and used military regulations to discipline the labor force.[27] Every workshop had one to three managers.[28] Under the chief work supervisor were procurement agents. They were stationed in Hong Kong and Burma, purchasing shipbuilding materials. All these junior managers bore the title of special delegate (*weiyuan*).

In 1873, the Fuzhou Navy Yard had 130 Chinese administrators.[29] In 1875, the number might have increased to 197.[30]

The Workforce

The navy yard recruited Chinese workers, but not all of them were unskilled in Western engineering. In late 1867, Giquel went to Shanghai to recruit 129 sailors and workers (referred to as "blacksmiths from Jiangxu and Zhejiang" in Chinese documents).[31] It is likely that they had experience of working in foreign-owned machine shops in Shanghai, as suggested by the British diplomat who reported on the development of the navy yard.[32] By August 1868, it employed around 2000 to 3000 workers. Among them around 600 worked in the dockyard and 800 in the workshops. Others were laborers (referred to as "coolies" in English documents) and 500 soldiers guarding the navy yard.[33] In 1874, there were 500 wood workers (carpenters, cabinet makers, and molders) and 600 iron workers.[34]

In every workshop, workers were managed by the foreign foreman who had the disciplinary power. The chime from the tower clock in front of the director' office regulated the workers' hours. After work, the workers had to go back to their living quarters. All movements were prohibited in the evening. Drinking, gambling, opium-smoking, and unruliness were forbidden.[35] All the workers had to undergo a training program that qualified them as modern shipbuilding workers by teaching them a foreign language, engineering skills, technical drawings, scientific knowledge, and mathematics.

Training on the Job

Since most of the Chinese workers recruited by the navy yard did not have the skills to operate machine tools, the foreign technicians immediately started to organize the Chinese workers into different specialties, teaching them Western skills in making sand and wooden molds, iron and copper foundering, and machine tools operation.[36] Moreover, Giquel considered skilled Chinese workers who had already reached their middle age to be mentally worn-out and therefore unable to learn new skills. Hence, he set up the School of Apprentices (*Yipu*) by October 1867, recruiting young workers between the ages of 15 and 18 *sui*.[37] In the School of Apprentices, the foreign teachers taught the students French, arithmetic, scientific and engineering knowledge. These students were expected to become foremen in the workshops.[38]

Apart from engineering skills, learning to read and draw technical drawings was necessary for modern workers to understand the machine designer's requirement and make the machine parts accordingly, which represented a major departure from China's own technological tradition. Hence, in January 1868, Giquel introduced the skill of reading ship plans to the Chinese workers. Giquel instructed the foreign draftsmen to draw a detailed full-scale 150 horsepower ship plan on the floor annotated with dimensions, to teach the Chinese carpenters to read the plan and build the hull accordingly.[39]

Furthermore, Giquel made the foreign draftsmen teach the Chinese workers to produce drawings. In 1868, he set up a Drawing Office (*Huaguan* or *Huishi yuan*). Not only did the Drawing Office turn out all the ship plans and drawings of machine parts for the workshops, it also functioned as a school. Giquel recruited young Chinese workers who were talented in drawing into its two courses. One course taught drafting ship plans and the other technical drawing.[40]

By teaching the skills of drafting, operating machine tools, making molds, and metal smelting and casting, Giquel and his technicians implanted the modern manufacturing methods in the Fuzhou Navy Yard. The production procedure started in the Drawing Office, where the draftsmen produced the plans, which were to be sent to the pattern shop for making molds. The foundry used these molds to cast and finish the components, and then sent them to the fitting-up shop for assembling.[41]

The result of the training was the first steamboat, the *Wannian Qing*, which was tested at sea in September 1869.[42] It was fitted with a purchased 150-horsepower engine and a screw propeller system.[43] It was 1450 tons and its dimension was 223 feet long, 29.5 feet wide, and 13.2 feet draft. It was armed with six purchased 14 cm smoothbore muzzle loading guns. By the end of 1870, four steamboats of the same class were built. (See Table 5.1 for more details about the Fuzhou-built steam vessels).

However, these four boats were fitted with purchased engines.[44] To fulfill the ambition of building the steam engine, the navy yard started to build a 150 horsepower engine in the winter of 1869 by copying the parts of a purchased engine. The procedure was similar to what we have discussed above. First of all, the Drawing Office measured the dimensions of the parts of a whole engine and drew their plans on a reduced scale. The plans were sent to the pattern shop for making molds, and then the molds were used by the foundry to cast and forge the components. After the components were machined in the fitting-up shop, the boiler shop produced the copper tubes of various descriptions to connect the engine parts.[45]

In June 1871, the first Fuzhou-built steam engine and its boiler were completed. It was fitted on the fifth steamboat, the *Anlan*, and ran successfully. By the end of 1876, the Fuzhou Navy Yard had built eleven 150-horsepower engines.[46]

At this early stage, copying seemed to be a practical way of making the workers practice the skills they were taught and to understand the mechanism of the steam engine. After they were familiarized with the skills, the

Table 5.1 Steamship built in the Fuzhou Navy Yard, 1868–1895

	Name	Description	Engine (Nominal H.P.)	Displacement (Tons)	Work begun (keel laid down)	Launched	Trial run
1	Wannianqing	Wooden transport	150	1450	18 Jan 1868	10 Jun 1869	18 Sept 1869
2	Meiyun	Wooden gunboat	80	515	8 Feb 1869	6 Dec. 1869	9 Jan, 1870
3	Fuxing	Wooden gunboat	80	515	6 Dec. 1869	30 May 1860	Sept/Oct 1870
4	Fubo	Wooden transport	150	1258	Ca. Jun 1860	22 Dec. 1870	1 Apr 1870
5	Anlan	Wooden transport	150ª	1005	26 Nov 1870	18 Jun 1871	Ca. Dec. 1871
6	Zhenhai	Wooden gunboat	80	572	29 Mar 1871	28 Nov 1871	July/Aug 1872
7	Yangwu	Wooden corvette	250	1393	12 Jul 1871	23 Apr 1872	Dec 1872
8	Feiyun	Transport	150ª	1285	Aug 1871	3 Jun 1872	Oct 1872
9	Jingyuan	Gunboat	80	572	1 Dec 1871	21 Aug 1872	26 Mar 1873
10	Zhenwei	Gunboat	80	572	24 Jun 1872	11 Dec 1872	17 Aug 1873
11	Ji'an	Transport	150ª	1285	25 Jul 1872	2 Jan 1873	27 Sept 1873
12	Yongbao	Transport	150ª	1391	23 Oct 1872	10 Aug 1873	19 Oct 1873
13	Haijing	Transport	150ª	1391	28 Feb 1873	8 Nov 1873	Jan/Feb 1874
14	Chenghang	Transport	150ª	1391	3 Jul 1873	16 Jan 1874	Mar/Apr 1874
15	Taya	Transport	150ª	1391	16 Aug 1873	16 May 1874	Aug/Sep 1874
16	Yuankai	Transport	150ª	1285	5 Dec 1874	4 Jun 1875	Jul 1875
17	Yixin	Transport	50ª	Unknown	4 June 1875	28 Mar 1876	10 Jul 1876

(*continued*)

5 TRAINING WORKERS AND ENGINEERS: THE FUZHOU NAVY YARD, 1866–1895

Table 5.1 (continued)

Name	Description	Engine (Nominal H.P.)	Displacement (Tons)	Work begun (keel laid down)	Launched	Trial run
18 Deng yingzhou	Transport	150[a]	1285	21 Jul 1875	Jun/Jul 1876	15 Sep 1876
19 Tai'an	Transport	150[a]	1285	12 Jul 1875	2 Dec 1876	3 May 1877
20 Weiyuan	Composite sloop	750	1200	2 Sep 1876	15 May 1877	14 Sep 1877
21 Chaowu	Composite sloop	750[a]	1200	2 Dec 1876	19 Jun 1878	21 Sep 1878
22 Kangji	Composite merchant	750[a]	1200	12 Jul 1878	20 Sep 1879	8 Feb 1880
23 Chengqing	Composite sloop	750[a]	1200	25 Jul 1879	22 Oct 1880	29 Dec 1880
24 Kaiji	Composite cruiser	2400	2200	9 Nov 1881	11 Jan 1882	29 Sep 1883
25 Henghai	Composite sloop	750[a]	1200	17 May 1883	18 Dec 1884	15 Jul 1885
26 Jingqing	Composite cruiser	2400	2200	Jan 1884	23 Dec 1885	11 Aug 1886
27 Huantai	Composite cruiser	2400	2200	Jan 1884	15 Oct 1886	30 Aug 1887
28 Guangjia	Composite sloop	1600	1300	24 Nov 1885	6 Aug 1887	4 Dec 1887
29 Pingyuan	Steel-hulled cruiser	2400[a]	2100	7 Dec 1886	29 Jan 1888	27 May 1888
30 Guangyi	Steel-hulled cruiser	2400	1000	2 Jan 1887	28 Aug 1889	30 Nov 1890
31 Guangbing	Steel-hulled cruiser	2400	1000	18 Jul 1887	11 April 1891	18 Dec 1891
32 Guanggeng	Composite gunboat	400	320	1 Feb 1888	30 May 1889	30 Oct 1889

(continued)

Table 5.1 (continued)

Name	Description	Engine (Nominal H.P.)	Displacement (Tons)	Work begun (keel laid down)	Launched	Trial run
33 Guangding (Fujing)	Steel-hulled cruiser	2400	1000	4 Nov 1889	20 Jan 1893	16 Dec 1893
34 Jianjin (Tongji)	Steel-hulled training ship	1600	1900	9 Jan 1894	12 Apr 1895	15 Sep 1896
35 Fu'an	Steel-hulled merchant	750	1700	2 Dec 1895	16 Aug 1897	Unknown

Source: David Pong, *Shen Pao-chen*, p. 219; *Conway's All The World's Fighting Ships 1860–1905* (London: Conway Maritime Press, 1979), pp. 395–440; Wang Zhiyi, *Zhongguo jindai zhaochuan shi* (Beijing: Haiyang chubanshe, 1986), p. 99, pp. 110–111; and data compiled from *Yangwu yundong*, vol. 5 (Beijing: Shanghai: Shanghai renmin chubanshe 1961)

[a] Steam engines built in the Fuzhou shipyard

workers would be able to reproduce similar engines without much difficulty. Yet, the ambition of the five-year contract was to train engineers who were able to design engines and ship hulls. Such training would have to be done through engineering schools that were to be set up, employing a curriculum that provided more than engineering skills and drawing. In February 1867 the Fujian officials had conducted an examination to recruit student into the Naval School.[47]

THE NAVAL SCHOOL

The Fuzhou Naval School was China's first modern professional school, which would be turning out China's first generation of modern naval officers and naval engineers. The school was divided into the English and the

5 TRAINING WORKERS AND ENGINEERS: THE FUZHOU NAVY YARD, 1866–1895

French Division. The former was the School of Navigation (*Hou xuetang*, literally the rear school). The latter was the School of Naval Construction (*Qian xuetang*, literally the front school). There were also other schools, the School of Design (*Huishi yuan*), and the School of Apprentices (*Yipu*).[48]

The School of Navigation was to train navigators, onboard engineers, and naval officers. The courses were taught in English. The curriculum included English, geography, arithmetic, algebra, geometry, trigonometry, navigation, and nautical astronomy. By 1871, 73 students had graduated. In September 1873, the school admitted another nine students.[49] The engineering class recruited students who had experience in working on iron material from Shanghai and Hong Kong.[50] I shall not discuss the training provided by this school in detail because the technology of navigation is not the focus of this book.

In the School of Naval Construction, all courses were taught in French, because there were no textbooks in Chinese, nor could the French teachers and technicians speak Chinese. Hence the students had to start from learning French. The program was designed to enable Chinese students, who had no knowledge or skills of steam technology before entering the school, to build the steamers set by the five-year contract. Hence, it contained both theoretical and practical trainings.

In terms of theoretical training, the branches of mathematics and physics. Arithmetic and geometry was necessary for students to understand and calculate the functions and the dimensions of engine parts and ship hulls, so they could design and reproduce different parts of steamships. To reproduce that engine and ship parts according to ship plans, it was necessary for students to learn descriptive geometry, which was the theoretical basis of technical drawings. Students also had to learn physics, which would enable them to understand the pressure, gravity, heat, and other physical phenomena sustained by engines and ships hulls. Furthermore, to understand the natural forces, the students had to learn statics and mechanics, in which trigonometry, analytical geometry, the infinitesimal calculus would be necessary for them to form general formulae applicable to all the details of construction.[51]

The students' theoretical knowledge of shipbuilding would later be put into practice. From 1874 they were required to engage in the works or engine building and hull construction under the instructions of the French engineers in the workshops. Then only could the students get familiar with the work of various machine shops and thus become capable of directing the workmen.[52]

The School of Naval Construction started with only 10 to 12 students and one French teacher. By April 1868, the number had increased to 40 students and two teachers of physics, chemistry, and mathematics.[53] From 1871, the students were required to take courses in practical hull construction and machinery operation in the workshops under the instruction of the foreign technicians.[54] During the five-year term, 105 students were admitted. But in 1874, only 39 students remained in the school.[55]

The School of Design, which was attached to the Drawing Office, had the responsibility of training draftsmen. Its curriculum included French, arithmetic, geometry, descriptive geometry, calculus, perspective, and a complete course on the 150 horsepower marine engine.[56] The students were required to reproduce working plans of all the parts of the steam engine. They were also required to spend some hours daily in the workshops to be familiar with engineering skills.[57] Giquel selected 10 students to attend the School of Naval Construction for further study in April 1871.[58] In 1872, 30 students were in the Drawing Office, but in November 1873, when Giquel delivered his final report to Shen Baozhen, 22 students remained.[59]

The School of Apprentices trained apprentices engineering skills and technical drawings, and also French, arithmetic, geometry, descriptive geometry, algebra, design, and the mechanism of the steam engine.[60] In November 1873, there were 77 apprentices in the school.

We cannot overemphasize the significance of the School of Naval Construction. It was China's first engineering school that systematically imported from the West a technology from its scientific principles to the engineering application. It was attached to a navy yard that enabled its students to practice their theoretical knowledge in the near future. Compared to other new educational institutions in late Qing China that introduced Western knowledge in the period between the 1860s and the 1880s, the schools of the Fuzhou Navy Yard were much more technical. The School of Combined Learning (*Tongwen guan*) in Beijing, and the School for the Diffusion of Languages (*Guangfanyan guan*) and the Translation Department (*Fanyi guan*) that were attached to the Jiangnan Arsenal in Shanghai aimed at training translators or diplomats and the syllabus of scientific subjects only provided general knowledge on science.[61] In 1867, the *Zongli yamen* added a department of mathematics and astronomy into the *Tongwen guan*, which was set up in 1861 for training interpreters and translators. Yet, the number of its students was no more than five. As shown in Chap. 3, the contents of the teaching were basic and simple, although the curriculum was reorganized in 1869 and was introduced theoretical and practical mechanics, and differential and integral calculus. No engineers were expected to be trained by this school.[62]

The School for the Diffusion of Languages was originally established by Li Hongzhang in Shanghai in 1863 and was incorporated into the Jiangnan Arsenal in November 1869. It began as a language school for training translators, although the curriculum included mathematics and physics.[63] It aimed to train translators and interpreters.

As discussed in Chap. 4, the Translation Department was established by Zeng Guofan in 1868 for introducing Western knowledge to Chinese readers. It recruited Western missionaries to translate books in Western languages into Chinese with the help of Chinese staff. The subjects ranged from Western political systems, military tactics, railways, marine navigation, mathematics, mining, metallurgy, and gunpowder, to marine steam engine management. Yet, the books were not selected to be part of the curriculum of any engineer training program but aimed at the general public. Nor did the Department run any such courses.

Therefore, the graduates of the School of Naval Construction were to be China's first and only group of domestically-trained engineers. They were to design engines and ships in the 1880s and later participated in railway building in the 1890s and 1900s. No other new educational institutions could match such professionalism in technology until the Jiangnan Arsenal set up an engineering school, the School of Technology (*Gongyi xuetang*) in 1898 that taught engineering knowledge and skills, technical drawing, and scientific subjects of mathematics, physics, and chemistry.[64]

However, although the training program of the school of naval construction might seem to have been complete, both the theoretical and practical training was not enough to enable the students to enter the workshops to direct shipbuilding works. Giquel explained to Shen Baozhen that because an engineer had to know the steam engine well and to organize a factory, it was necessary for the students to study various types of engines and their drawings as well as the works of different workshops. Most importantly, they had to engage in the practical works of building different types of engines. Nevertheless, the Fuzhou Navy Yard and the five-year contract could not provide such a training program, nor was China's engineering industry sufficient to form such an environment. Therefore, Giquel suggested, it would be necessary for the students to go to Europe to undergo a four-year training in order to acquire the necessary experience. By learning the most advanced shipbuilding technology there, Giquel argued, the students would be capable of designing all the various parts of an engine. Such are the acquirements that are demanded in industrial pursuits in the office of design.[65]

Giquel's suggestion shows that towards the expiry of the contract, Giquel and Shen Baozhen were contemplating the possibility of extending the training program of naval engineers to Europe.

THE NEXT STEP

By 1872, the Fuzhou Navy Yard was turning out two steamboats a year, and its workers and students were being trained. It seemed to have been fulfilling the Qing government's ambition in improving maritime defense. However, with its budget overrun and the effectiveness of those steamers caused concern. In January 1872, Song Jin, member of the Grand Secretariat, criticized the Fuzhou Navy Yard for these two problems. He suggested that the shipbuilding program should be stopped and the steamers be rented to merchants.[66] The imperial court allowed the Fuzhou Navy Yard to carry on after Zuo Zongtang, Shen Baozhen, Li Hongzhang, and the *Zongli yamen* strongly defended the necessity of the shipbuilding program.[67]

However, Song Jin's criticism still created immense pressure on Shen Baozhen on the issue of spending. Hence, before the end of the five-year term in 1874, Shen had decided to reduce the number of foreign technicians to avoid further political controversy and at the same time to decrease the spending on foreign technicians amounting to more than 100,000 taels a year, which was one sixth of the Fuzhou Navy Yard's annual budget.[68] From hindsight, dismissing foreign technicians in such a critical moment was not a wise decision. The navy yard's workers had just gone through a five-year training program of building basic design of steamers. Engineering students were still in the naval construction school and could hardly work as independent engineers. Besides, shipbuilding technology in the West had advanced far beyond the works of the five-year contract. Nevertheless, Shen and Giquel were working in a political atmosphere in which impatient officials did not see the necessity of time and money in transferring technology. They were under immense pressure to produce good result with limited funds. Hence they had to think about curtailing spending at the same time consider the question of how to keep up the Fuzhou Navy Yard's technology with Western developments, which were iron hull construction and the compound engine.

Iron and the Compound Engine

Until the mid-nineteenth century, wood was the dominating material in shipbuilding. Iron was not widely adopted in hull construction because of the problems of its brittleness, corrosion, and its potential effect on the compass. Besides, the iron hull at the time was more difficult to repair than the wooden one.[69] However, wood had its limitations. Shipyards had to stock a large amount of wood to ensure that the building works were carried on smoothly. Yet, one third of wood's weight would be lost in the process of planing. Secondly, the length limit of the wooden hull was 300 feet. Thirdly, wooden hull's elasticity was liable to "hogging and sagging" in the seaway, which caused difficulty in aligning the shaft of the screw propeller. Also, the screw propeller caused stress to the stern and, therefore, led to leaks. Furthermore, the engine's boiler increased fire risk to the wooden hull.

In contrast, iron did not lead to these defects. Iron increased strength, solidity, and the durability of the hull. It diminished fire risk, reduced hull weight, increased cargo space, and facilitated water-tight subdivision. It made the building of much larger vessels possible and prevented leaks in the stern. With the progress of iron foundering techniques, it made the hull more economical in construction and repairs. Generally speaking, iron made it possible to decrease both the weight of the vessel and the thickness of the hull, allowing increases of up to 35 percent in deadweight cargo capacity and of 20 to 50 percent in the hold space in proportion to the exterior dimensions.[70] Therefore, after the 1850s, iron was more widely adopted in shipbuilding, and after the Crimean War and the American Civil War, newly-built warships were all iron-hulled or at least armored wood-hulled with iron frame—composite construction.

As for marine engineering, the compound engine was becoming the norm. Since the 1840s, the simple flue boiler was being replaced by the tubular boiler, in which the gas tubes significantly increased the heating surface and accelerated the speed of generating steam pressure with less fuel. The steam pressure was increased from 10 to 15 lbs. per square inch above the atmosphere. Besides, the boiler was lighter and more compact. In the 1860s, the new boiler could generate steam pressure of up to 60 lbs. per square inch. The high pressure helped the materialization of the idea of compounding. This meant that the steam could expand in two stages in two separate cylinders, one of them smaller for higher pressure the other larger for lower pressure. Such a mechanism greatly increased the economy of the steam and reduced the consumption of coal. In other words, ships could carry less coal but more cargo, or, in the case of warships, heavier armor, more guns, shells, and seamen.[71]

During the five-year contract of the Fuzhou Navy Yard such technologies were not introduced for one simple reason: the workers had not yet gone through the basic training. Yet, Shen Baozhen and Giquel could not have ignored the developments in the West. According to the British diplomat's report, Giquel had suggested building a composite with a compound engine in 1873. Shen did not agree to the proposal.[72] It is possible that his priority was to complete the five-year contract. Moreover, if Shen Baozhen and Giquel wanted to import newer technology for iron construction and the compound engine, they had to come up with a way that would involve less foreign presence in the navy yard and let the students of the naval construction carry out more training. Yet, laying-off most of the foreign technicians might not have been beneficial to China's learning the new technology.

After Giquel's contract expired in January 1874, Shen retained him, his deputy, the chief engineer Jouvet, and the two teachers of the School of Naval Construction.[73] Besides, in January 1874 he memorialized to allow the navy yard to carry on building steam vessels and proposed to send the Fuzhou students to embark upon further training in Europe and to bring back "the principle of the new technology."[74]

Yet, before any decision was made, in May 1874, Japan invaded Taiwan and created a military crisis. Senior Chinese officials found out that Japan had two ironclads and, hence, called for the acquisition of new type of warships.[75] (The impact of the ironclad on the navy yard's finance will be discussed in Chap. 6.) Senior officials' call for the acquisition of ironclads made Shen Baozhen's plan easier to push through. He proposed to the imperial court that the Fuzhou Navy Yard should learn the composite construction and the compound engine in order to build ironclads in the future.[76]

In January 1875, Shen Baozhen finally obtained court permission to import the new technologies.[77] Before the compound engine and the iron frame arrived at Fuzhou, Shen Baozhen was appointed governor-general of Jiangsu, Jiangxi, and Anhui in May 1875, and his successors implementing the new plan he had laid down.[78]

The training and building works of the new technologies started in July 1876.[79] Yet, the progress was slow.[80] Although the available documents do not tell us more details, the delay could have had to do with the difficulties of building the more sophisticated boiler and the demand for producing the iron parts of the frame. The much-reduced number of foreign technicians may have contributed to the slowness in the learning process. To learn the new technology, the navy yard hired seven foreign technicians from Britain and France to teach the skills for making copper tubes and iron plates. It also purchased a more advanced iron rolling mill and a copper rolling mill.[81] The first composite frigate, the *Weiyuan* (dimension 210 × 29.5 × 12.5 ft, 1200 tons), which was fitted with a 750-horsepower compound engine, was launched in May 1877. It was armed with one 7 inch and one 4.7 inch muzzle-loading guns. By December 1880, the navy yard built four frigates of this class.[82]

The Training Program in France and in Fuzhou

When the navy yard was carrying out the plan of introducing iron structure and the compound engine, Giquel and Shen Baozhen were also pushing their plan to send the Fuzhou students to Europe for further training. In March 1875, Shen Baozhen selected three students of the French Division including Wei Han and Chen Zhao'ao, who were to become the leading engineer in the Fuzhou Navy Yard in the 1880s and 90s, and two students from the English Division, to go to Europe with Giquel.[83] Giquel took them to visit the navy yards in Britain and France. In June 1876, when Giquel came back to Fuzhou with the compound engines, Wei Han, Chen Zhao'ao, and the other students were left in France for further training.

In January 1877, Li Hongzhang and Shen Baozhen memorialized and obtained permission to send 30 Fuzhou students, including 12 students and 6 apprentices from the French Division and 12 students from the English Division, to study in Europe for a maximum of three years.[84] They set out in March 1877 with two supervisors including Giquel.[85] The French-division students and apprentices went to France, and the English Division students to Britain.[86] In France, students were divided into three professions: naval architecture, marine engineering, as well as mining and metallurgy. They entered reputable education institutions. Wei Han joined two other students to enter the École de Construction Navale in Cherbourg to learn naval architecture. Five students, including Chen Zhao'ao, entered the Toulon Navy Yard to learn naval engineering. Four went to the Le Creusot ironworks and one to the École Nationale Supérieure des Mines in Saint-Étienne to learn mining and metallurgy.[87] In the following year, these five metallurgy students were transferred to École Nationale Supérieure des Mines in Paris. In October 1877 the Fuzhou Navy Yard sent five more apprentices to France for training.[88]

Apart from studying, the students helped to establish a direct link with the institutions in which they studied. Their supervisors had the task of investigating the latest technology of hull design, engines, firearms, and other machinery. They obtained the drawings of new technologies and requiring the naval construction students to study them. After that, the students sent the plans back to Fuzhou.[89] Moreover, the naval construction students were required to have the ability to build the latest model warships and all their components without foreign assistance.[90]

In November 1879, Li Hongzhang and Shen Baozhen memorialized to the throne, suggesting that the Fuzhou Navy Yard should send another group of Fuzhou students abroad.[91] However, it took almost three years for the navy yard to select students to form the second group. The reason for the delay might have to do with the death of Shen Baozhen in December 1879 and Giquel's departure from Fuzhou in 1875. In 1882, the second group, including eight from the French Division and six from the English Division, left for Europe. The French Division students were not studying shipbuilding but other aspects of military technology. Five of the students studied in France. Two of them studied fort construction, and three were trained in ordnance, gunpowder and other explosives technology. One went to Germany to learn the technology of underwater mines and torpedoes. They returned to Fuzhou in 1886.[92]

Those students were to bring new technologies back to Fuzhou. Among them, the eight naval construction students from the first group made the most important impact on the Fuzhou Navy Yard after they completed the training in September 1880. They were to manage technical matters in the Fuzhou Navy Yard, and the first challenge was the cruiser project, which was initiated in 1876.

Catching Up

Giquel had long understood that the Fuzhou vessels were not up to the Western standard. In September 1876, he earned Li Hongzhang's support and made a request to the imperial court for permission to build cruisers.[93] Li considered that cruisers could fill the strategic gap between gunboats and not-yet-purchased ironclads. But the project was not carried out until Zuo Zongtang, appointed governor-general of Jiangsu, Jiangxi, and Anhui in 1881, agreed to pay half of the building costs.[94]

Giquel's proposed cruiser was 2200 tons and its engine was a 2400-horsepower triple expansion compound engine with eight boilers generating 68lb steam pressure.[95] It was 252 feet long, 36 feet wide, and 18.5 feet draft. It could cruise up to 15 knots. Although the size was not dramatically larger than the 750-horsepower composites, the demand for marine engineering was much higher. Although the navy yard had upgraded part of the facilities since the composite project, around 60 to 70 percent of the components, especially the steel parts, were imported.[96]

The first cruiser, the *Kaiji*, was completed in September 1883 and it attracted senior officials' attention. Zuo Zongtang ordered two more.[97] While the building work was being carried out, the Sino-French War broke out and, in August 1884, in the Battle of the River Min, the invading French fleet damaged the workshops. The damage affected the shipbuilding work. Hence, many of the iron parts had to be purchased from Britain.[98] In December 1885 and October 1886, the second and third cruisers were launched respectively. They were both based on the same design as the *Kaiji* but with modifications for fitting more guns and two torpedo launchers.[99]

The three cruisers might have been impressive, compared with earlier smaller Fuzhou vessels. Yet, they did not match the Western standards. The newly-built cruisers in Western Europe were up to 9000 tons, 400 feet long, and could attain a speed of 22 knots. Triple expansion engines of 8000 horsepower were common. Furthermore, steel had become the major material of hull construction.[100]

STEEL

Steel, a metal alloy whose major component is iron with up to 2 percent of carbon, is less brittle than cast iron (carbon content higher than 2 percent) but stronger than wrought iron (or pure iron, whose carbon content is less than 0.1 per cent). Steel making demands a careful control of the amount of carbon by burning it out of cast iron or adding it into wrought iron. By the mid-nineteenth century, it was time-consuming, laborious, unreliable, and, therefore, expensive. Yet, after the introduction of the Bessemer converter and the Siemens-Martin process, introduced in 1856 and 1865 respectively, steel became a widely-used and favorable industrial material because of its strength and light weight. In shipbuilding, its use could reduce the thickness of iron by one-fourth, which is to a twentieth of an inch.[101] Hence, since the late 1870s, shipbuilding in the West was moving towards steel hulls and, in warship building, towards steel armoring.[102]

The Fuzhou Navy Yard's engineers would have known this. Some of them had undergone the training in mining and metallurgy in France in the 1870s.[103] Some of them were sent by Li Hongzhang to Germany to observe the construction of two steel-hulled battleships and one steel-protected cruiser in the 1880s.[104] During their two-year stay at the German shipbuilder, Vulcan, they witnessed the advancement of German shipbuilding technology.[105]

However, using steel in hull construction posed a challenge to the navy yard's workforce and metallurgical capacity. Carpenters would not have been needed in building the hull. Unlike wooden planks that could be bent against the frame, iron- or steel-hull components had to be made with geometrical precision before they could be riveted into position. Consequently, iron or steel workers had to replace carpenters. Machines also had to be changed. Mechanical saws and other carpentry tools would be useless and the machines of shearing, punching, and drilling iron- and steel-plates had to be brought in. The skills of iron and steel working and machines of riveting would have to be used extensively rather than being limited to the building of machinery and the iron frame.[106]

In 1885, the Fuzhou Navy Yard obtained the court's permission to build a twin-engine steel-hulled ship. It was based on a French design.[107] The construction work started in December 1886. Although, by 1886, the Navy Yard might have had a certain degree of steel making capability, the scale must have been limited.[108] Hence the navy yard had to import most of the steel parts including the hull, the armor, the boiler, and the keel. Yet, the engine was built within the navy yard.[109] In May 1888 the ship, *Longwei*, was completed. It was 2100 tons and was equipped with two sets of 1200 horsepower triple expansion engines that enabled it to cruise to 14 knots. It adopted the forced draft design to increase the efficiency of the boiler furnace. It was 215 feet long, and 40 feet wide, and 16 feet draft. Its hull had an 8-inch thick above-waterline belt-armor and the turret armor was 14 inches thick. It also had one 11-inch revolving Krupp gun in an armored turret on the fore deck, which was worked by hydraulic machinery. Behind the gun was an armored tower with telephonic communication to all parts of the ship. It was also equipped with two torpedo launchers. The specifications made it a formidable ship in China. Li

5 TRAINING WORKERS AND ENGINEERS: THE FUZHOU NAVY YARD, 1866–1895

Hongzhang immediately took it over and renamed it as *Pingyuan* (Fig. 5.1).[110]

After the *Pingyuan*, although the navy yard still built composites, steel-hulled vessels became its focus. By co-operating with Zhang Zidong, governor-general of Guangdong and Guangxi, who funded half of the building costs, it started to build three vessels, two of which were steel-hulled cruisers and one was a shallow draft composite gunboat.[111] The two cruisers, *Guangyi* and *Guangbing* (dimensions 235 × 27 × 13 feet) had steel protected decks but the frames might have been made of iron.[112] They were 1000 tons and equipped with 2400 horsepower compound engines and could cruise at 16.5 knots. They were armed with guns and torpedoes. The composite gunboat, *Guanggeng*, (144 × 20 × 10 ft) was 320 tons and was equipped with a 400 horsepower engine of unknown type. All these ships show that the navy yard had upgraded its technical capabilities and facilities to a certain level in order to meet the demand of building more complex vessels.

Fig. 5.1 The *Pingyuan* after it was captured by the Japanese Imperial Navy. Source: Richard N.J. Wright, *The Chinese Steam Navy, 1862–1945* (London: Chatham Publishing, 2000), p. 81

Facilities Upgrade

With the growing complexity of the vessels built after the mid-1870s, the Fuzhou Navy Yard kept adding new machinery such as steam engines, boilers, furnaces, cranes, rolling mills, planers, boring machines, and other machine tools to its workshops.[113] By July 1879, heavy and large parts such as the iron plates as large as 10 square feet and shafts of 75 inches in diameter were produced in the navy yard.[114] A 7000-pound steam hammer was purchased for making shafts. Machinery was purchased for making the large engine cylinders.[115] Two steam engines, one 40-horsepower and the other 25-horsepower, were produced to provide power for the workshops.[116] Therefore, the cylinder, the surface condenser, the boiler and more than a thousand other parts were produced in the navy yard. However, the Fuzhou Navy Yard never acquired adequate steel-making capacities. Steel parts such as the propeller shaft and the crankshaft had to be ordered from Britain.[117] Yet, after 1886, building the steel-hulled cruiser still pushed the navy yard to bring in steel handling machinery such as a steel rolling mill, planer, shearers, and punchers.[118] However, the available documents do not suggest that there was any large-scale expansion, although they show that in 1883 the Drawing Office was expanded and, in 1886, a new warehouse was added, slips were refurbished and furnaces in the ironworks were enlarged.[119] A major addition was the dry dock that was started by around January 1888 but was only completed in 1893.[120]

As mentioned above, the change of technology would have brought about the change of workforce. Hence the organization of the navy yard would unavoidably be altered.

Change of Organization

The new technologies introduced into the Fuzhou Navy Yard in the 1870s and the 1880s were institutionalized, thanks to the organizational foundation laid down by the foreign technicians in the five-year contract, the organization was to undergo some changes after the end of the contract.

After Shen Baozhen left Fuzhou, the navy yard was directed by less prestigious officials. After the end of the first five-year contract in January 1874, the organization went through a major change. There were no longer foreign supervisors directing shipbuilding works. The few retained or newly-hired technicians were under short-term contracts for approximately a year.[121] Shen Baozhen appointed Giquel the supervisor of the Schools of Naval Construction and Navigation. The former chief engineer was moved to supervise shell production to meet the needs of Taiwan's defense and the building works of the compound engine.[122] After 1875, Giquel was no longer involved in managing either the navy yard or the schools. There were no foreign technicians after 1880 when the students of naval construction who studied in France obtained the license of engineer and came back to Fuzhou.

The returned naval construction engineers formed the department of engineering and construction (*Gongcheng chu*).[123] Although available documents do not reveal much about the department, it seems to have managed all the technical works of the navy yard. The engineers were granted official or military titles such as "expectant vice-magistrate on priority list" (*jinxian xuanyong xiancheng*) or "expectant brigade vice-commander on priority list" (*jinxian dusi*).[124] In the early 1890s, their titles equaled to those of expectant circuit intendant (*daotai*). They were considered to be low or middle-ranking officials in Qing officialdom. Yet, the Fuzhou Navy Yard did not create formal positions for them. They were referred to as "students" (*xuesheng*) or "students who had studied abroad" (*chuyang xuesheng*) in Chinese documents.

The Training Program in Europe and Fuzhou

As mentioned above, the second group of the French Division students who were trained in Europe did not focus their study on shipbuilding. This might have to do with Li Hongzhang's interest in building up his *Beiyang* fleet and obtaining different aspects of technology, especially the railway. Li Hongzhang's influence is obvious in the navy yard's training program. Since the early 1870s, Li had maintained close links with the successive directors of the Fuzhou Navy Yard, discussing the matters of the training programs. Moreover, Li had been submitting proposals for sending Fuzhou students abroad, allocating funds for the program, reporting the students' progress in training, and memorializing to the throne for rewarding them.[125] Furthermore, after the death of Shen Baozhen, he continued to report on the matters concerning the training program abroad to the imperial court, and was the only senior official who supported the training program.[126] In November 1885, Li Hongzhang, not the director of the Fuzhou Navy Yard, obtained the court's permission for the third group of students to be sent to Europe.[127]

Hence, it is not surprising that the third group of students comprised of 10 cadets from Li Hongzhang's *Beiyang* fleet as well as Fuzhou's 10 English-division and 14 French-division students. In 1885, the 33 students went to Europe (one Fuzhou student was retained to serve in the *Beiyang* fleet). 13 of them went to France, and 12 finished their six-year course of study. Only two of them studied naval architecture and marine engineering, two studied mathematics and science relating to naval construction, two studied river control and bridge and railway construction, and six studied international law and French.[128] They returned to Fuzhou in 1891 but did not seem to have contributed much to the Fuzhou Navy Yard's technology throughout the 1890s.

As for training in Fuzhou, the number of the students in the French Division remained at the level of the 1870s.[129] In 1884, 106 students were in the French Division. 34 of them studied naval construction, 10 were in the School of Design, and 62 were in the School of Apprentice.[130] The French Division still kept two foreign teachers with the graduates who returned from Europe after 1880 as assistants.[131] The curriculum might have remained the same as under the first five-year contract. Yet, after 1878 the students of the French Division were required to learn English in order to be chosen either to be engine-room engineers or to be sent to Britain for further training.[132]

Nevertheless, the navy yard was fighting an uphill battle in keeping up with the development Western armament technology. In the 1890s, the change of naval tactics had given birth to the destroyer. Its high speed and armament of guns and torpedoes was designed to defend harbors and fleets against the threat from torpedo boats. Besides, the newly built capital battleships had reached the size of 15,000 tons and cruisers ranged between 2000 and 14,000 tons. In marine engineering, although the triple expansion engine that was introduced in the 1880s was still the norm, water tubes replaced traditional gas tubes in boiler construction. Such designs greatly improved fuel economy and reduced weight. The new engines could propel cruisers up to 20 to 23 knots and battleships 17 to 18 knots. The ordnance makers, helped by improved steel-making and machine tools technology, could build stronger and lighter guns that could be mounted on warships. These warships were the precursors of the Dreadnaught class of the 1900s.[133] Such battleships could reach the length of more than 520 feet and were equipped with ten 12-inch guns and 11-inch armor. They were equipped with state-of-the-art steam turbine that could push the enormous 18,000 tons vessels to a speed of 20 knots.

In contrast, in the Fuzhou navy yard, the shipbuilding work had begun to slow down in the late 1880s. The final project in the 1880s was the *Fujing*, which was of the same class as the *Guangyi*. The building work started in 1889 and dragged on for four years, until late 1893.[134] Similar projects would have normally taken around two years to complete in the 1880s. Besides, the navy yard had made few upgrades to the facilities since the late 1880s. Building the dry dock, the only major expansion work in the late 1880s, took six years to complete. Lack of funds was the main reason for such idleness. Chapter 6 will discuss how the Qing bureaucratic system that did not understand modern technology's demand for funds had contributed to the problem of funding the Fuzhou Navy Yard.

Conclusion

The Fuzhou Navy Yard started in 1866 with a five-year contract in order to learn how to build steam vessels. Its foreign technicians built a modern navy yard and started to train around 1400 men as modern shipbuilding workers, in addition to having 100 students in its schools. When the contract expired in 1874, navy yard workers were skilled at basic shipbuilding technology, and the students of naval construction had learned the principles of scientific knowledge that enabled them to embark on professional training in Europe. The achievement of the five-year contract paved the way for the technical progress of the 1870s and the 1880s. The navy yard introduced newer technology and, especially after the foreign-trained engineers returned in 1880, gradually began to catch up with the contemporary shipbuilding standards.

By 1894, the Fuzhou Navy Yard had built 33 steamers. 19 of them were wooden-hulled small-power steamboats. 10 were composite ships. The last four vessels were steel-hulled cruisers equipped with powerful and sophisticated compound engines. They were quality ships. In the Sino-Japanese War, eight Fuzhou-built ships joined the Battle of Yalü of 1894. Two steel-hulled cruisers, the *Pingyuan* and the *Guangbing*, were seized by the Japanese navy and later deployed in Russo-Japanese War in 1904.[135]

If we consider that in 1866 the majority of Fuzhou workers had no Western engineering skills and the newly-recruited students of the naval school could not understand French and mathematics, within just 23 years, the Fuzhou Navy Yard had made great technological progress. This shows how important a complete training program was that tackled the differences between Chinese and Western technology by introducing both scientific knowledge and applied skills. By the 1890s, the training program had produced around 29 engineers whose professions ranged from naval architecture, marine engineering, mining and metallurgy, gun making, and civil engineering.

The success of the Fuzhou Navy Yard in training engineers is attested by the continuation of these men's professional careers well into the twentieth century. Four of them remained in Fuzhou, four became the managers of other navy yards and arsenals in China. Five entered railway building. Seven or more joined the mining industry. Five became junior ministers when the Qing government launched a reform program and established the Department of Mail and Communication.[136] The most prominent of them was Wei Han and Chen Zhao'ao. They were among the first group of students who went to France, had been trained as a naval architect and marine engineer respectively. After 1880, they co-led the engineering department, directing shipbuilding works, especially the steel-hulled cruiser. Wei was promoted to chief engineer in 1897, vice-director in 1903, and for a year, in 1904, he became the director of the Fuzhou Navy Yard. In 1909 he was appointed general manager of the Guangzhou-Jiulong Railway.[137]

More importantly, the Fuzhou Navy Yard's achievement was not just the ships and the engineers, but the creation of an institution that could accumulate and practice knowledge and skills. By 1890, it already had the ability to select essential technology, master it, and reorganize it to meet the new technical demands. In the 1890s, although the navy yard still lagged far behind the fast advancing Western shipbuilding technology, it could have continued to catch up.

Yet, it had limits. As a leading shipyard in China, it did not have a better way of diffusing its technology other than the moving of limited number of engineers and technicians it had trained. In a more ideal situation, its school could have recruited more students that could find employment in private institutions. It might also published books or journals that diffused what the engineers had learned from Europe or developed within the navy yard to be known by engineers of other government or private institutions. Unfortunately, the directors and engineers did not seem to have had considered such options. Furthermore, as other Qing government undertakings, its development was restrained by the institutional setting, which will be discussed in Chap. 6.

NOTES

1. Thomas L. Kennedy, *The Arms of Kiangnan: Modernization in the Chinese Ordnance Industry, 1860–1895* (Boulder, Colorado: Westview Press, 1978).
2. Wang Xinzhong, "Fuzhou chuanchang zhi yange," *Qinghua xuebao* 8:1 (1932), pp. 1–57; Steven A. Leibo, *Transferring Technology to China: Prosper Giquel and the Self-strengthening Movement* (Berkeley: Institute of East Asian Studies, University of California, Berkeley, Center for Chinese Studies, 1985); Shen Chuanjing, *Fuzhou chuanzheng ju* (Chengdu, Sichuan remin chubanshe, 1987); Lin Chongyong, *Shen Baozhen yu Fuzhou chuanzheng* (Taipei: Lianjing chubanshe, 1987); David Pong, *Shen Pao-chen and China's Modernization in the Nineteenth Century* (Cambridge: Cambridge University Press, 1994); Basidi (Marianne Bastid) "Fuzhou chuanzheng ju de jishu yinjin," *Chuanshi yanjiu* No.10 (1996), pp. 104–114; Lin Qinyuan, *Fujian chuanzhengju shigao* (Fuzhou: Fujian Renmin chubanshe, 1999, revised edition).
3. Gideon Chen (Chen Qitian), *Tso Tsung-tang, Pioneer Promoter of the Modern Dockyard and the Woollen Mill in China* (New York, Paragon, 1961, reprint of the 1938 edition), p. 8.
4. Zuo Zongtang, *Zuo Wenxianggong quanji* (The Complete Writings of Zuo Zongtang), (Taipei: Wenhai chubanshe, 1979, reprint of a late nineteenth century edition), Memorials, 18:5b–6ab.
5. Gideon Chen, *Tso Tsung-tang*, p. 11.
6. The French naval forces in China had established a navy yard in Ningbo and built four gunboats. Chinese workers were hired and trained in this navy yard and local officials watched its development with interest. The French naval commander in China hoped to let the Chinese operate the navy yard on a permanent basis, but make it available for French needs. Therefore, the commander instructed Giquel and d'Aiguebelle to invite Zuo Zongtang to join the partnership. David Pong, *Shen Pao-chen*, p. 110; Steven Leibo, *Transferring Technology to China*, pp. 69–70.
7. Zuo Zongtang, *Zuo Wenxiang gong quanji*, Memorials, 18:5b–6a.
8. This account is based on Zuo Zongtang's June 1866 memorial, which was two years after the alleged West Lake steamship trial. *Zuo Wenxiang gong chuanji*, Memorials, 18:5b–6a; *Haifang dang* II, pp. 8–9. Steven Leibo and David Pong have pointed out that there are discrepancies between Zuo's accounts and Giquel's journal. Yet, the differences did not affect the initiation of the Fuzhou project. Steven A. Leibo, *Transferring Technology to China*, p. 70; David Pong, *Shen Pao-chen*, pp. 110–111.
9. *Haifang dang* II, p. 17, 20; *Yangwu yundong*, vol. 5, p. 10.

10. Zhongguo diyi lishi dang'an guan (ed.), *Xianfeng Tongzhi liangchao shangyudang*, volume 16. (Guilin: Guangxi shifan daxue, 1998), p. 216. The Moslem uprising, which began in 1862, had spread from eastern Shaansi to most of Xinjang.
11. Zuo Zongtang, *Zuo Wenxianggong chuanji*, Memorials, 20:67a–68a.
12. Zuo Zongtang, *Zuo Wenxianggong chuanji*, Memorials, 20:63ab; *Yangwu yundong*, vol. 5, pp. 24–25.
13. Steven A. Leibo, *Transferring Technology to China*, p. 7; David Pong, *Shen Pao-chen*, p. 179.
14. There is a large body of discussion on the French technical education, for example, Frederick Artz, *The Development of Technical Education in France, 1500–1850* (Cambridge, Mass.: The MIT Press, 1966); C.R. Day, *Education for the Industrial World: The Ecole d'Arts et Métiers and the Rise of French Industrial Engineering* (Cambridge, Mass.: The MIT Press, 1987); James M. Edmonson, *From Mecanicien to Ingenieur: Technical Education and the Machine Building Industry in Nineteenth Century France* (New York: Garland Publishing, 1986); Barton C. Hacker, "Engineering a New Order: Military Institutions, Technical Education, and the Rise of the Industrial State," *Technology and Culture* 34:1 (1993), pp. 1–27.
15. Zuo Zongtang, Zuo *Wenxianggong chuanji*, Memorials, 20:64b–66b; Prosper Giquel, *The Foochow Arsenal and Its Results: From the Commencement in 1867 to the End of the Foreign Directorate on the 16th February, 1874*, translated by H. Lang, (Shanghai: Shanghai Evening Courier, 1874), p. 17.
16. For more detailed discussion, see James Phinney Baxter, *The Introduction of the Ironclad Warship* (Cambridge, Mass.: Harvard University Press, 1933); Bernard Brodie, *Sea Power in the Machine Age* (Princeton: Princeton University Press, 1943).
17. *Haifangdang*, II, pp. 20–21. In July 1867, Shen Baozhen assumed the office of director of the Fuzhou Navy Yard. Wu Tang assumed the office of governor-general of Fujian and Zhejiang in early April 1867. He attempted to dismiss three Chinese directors who were appointed by Zuo Zongtang. Both Zuo Zongtang and Shen Baozhen memorialized the imperial court criticizing Wu Tang's acts. In January 1868 Wu Tang was appointed as governor-general of Sichuan and Ma Xinyi as governor-general of Fujian and Zhejiang. For more details of this incident, see David Pong, *Shen Pao-chen*, pp. 148–153; Lin Chongyong, *Shen Baozhen yu Fuzhou chuanzhen* (Taipei: Lianjin, 1987), pp. 295–307; Lin Qingyuan, *Fujian chuanzhenju shigao* (revised edition) (Fuzhou: Fujian renmin, 1999), pp. 40–42.

18. Huang Weixuan, "Fujian chuanzhen juchang gaocheng ji," in *Qiashantang tenggao*, quoted in Lin Qingyuan, *Fuzhou chuanzhengju shigao* (revised edition), p. 87. By December 1868, all the machinery that Giquel and d'Aigubelle ordered in France had arrived at Fuzhou. *Haifangdang*, II, p. 137.
19. Public Record Office, ADM 125/19, Prosper Giquel, "Memorandum on the Foochow Arsenal 19th Dec 1872."
20. *Haifangdang* II, p. 263; Prosper Giquel, *The Foochow Arsenal and its Results*, pp. 10–11. The British Consul in Fuzhou also knew Zuo Zongtang's ambition. He offered to let Zuo rent a British dock that belonged to Forster & Co, but Zuo refused. FO 228/408, Sinclair to Rutherford Alcock, 15 Sept 1866.
21. The *North China Herald*, 21 April 1870.
22. Sidney Pollard, *The British Shipbuilding Industry, 1870–1914* (Cambridge, Mass.: Harvard University Press, 1979), p. 111.
23. David Pong, *Shen Pao-chen*, p. 214.
24. Giquel recruited competent engineers to fill this position. The first chief engineer Adrien Marie Trasbot, was formerly an engineer of the Rochefort Arsenal, a leading shipyard in France. Trasbot left the Fuzhou Navy Yard in 1869 and another engineer Arnaudeau took over his position. D'Aiguebelle departed in 1870 and de Segonzac, a French naval officer who could speak Chinese, took over the position of deputy foreign directory. Historians have discussed in detail these disputes among the foreign technicians. They created unpleasant troubles for Giquel personally but not serious obstacles to the shipbuilding work. For more details, see Steven A. Leibo, *Transferring Technology to China*, pp. 88–106; David Pong, *Shen Pao-chen*, pp. 176–190; Li Chongyong, *Shen Baozhen*, pp. 319–330. After Trasbot left Fuzhou in 1869, two French engineers M. Jouvet, who had directed the office of design in one of the French engineering firms, and Gustave Alexandre Zédé joined the Fuzhou navy yard and became the chief engineers. He later became the director of naval construction at Cherbourg, another leading station of naval construction in France. *Haifangdang*, II, p. 471, Steven A. Leibo, *Transferring Technology to China*, p. 121.
25. Prosper Giquel, *The Foochow Arsenal*, p. 11.
26. The initial number of the Chinese directors was four but only three were serving in the navy yard. After 1872, the number dropped to one. For more details about these directors, see David Pong, *Shen Pao-chen*, pp. 162–176.
27. *Haifangdang*, II, p. 134.
28. Lin Qingyuan, *Fuzhou chuanzhenju* (revised edition), p. 104.
29. Prosper Giquel, *The Foochow Arsenal*, p. 14.

30. The number is from Shen Baozhen's memorial for rewarding the navy yard's administrators. *Haifangdang* II, pp. 557–575.
31. *Haifangdang*, II, p. 122.
32. Public Record Office, FO 17/787, "Report upon the Naval Arsenal at Foochow and the numbers of vessels of war belonging to the Chinese Navy stationed at that Port," Jamieson to Wade, 27th December 1878.
33. *Haifangdang*, II, p. 134. David Pong, *Shen Pao-chen*, p. 209. ADM 125/19, Prosper Giquel, "Memorandum on the Chinese Arsenals at Foochow," 19th December 1872.
34. Prosper Giquel, *The Foochow Arsenal*, p. 14.
35. *Haifangdang* II, p. 121; Lin Qingyuan, *Fuzhou chuanzhengju* (revised edition), pp. 86–87; David Pong, *Shen Pao-chen*, pp. 210–211.
36. *Haifangdang*, II, p. 134.
37. *Haifangdang*, II, pp. 134–135
38. Steven A. Leibo, *Transferring Technology to China*, p. 115.
39. *Haifangdang*, II, p. 113. The full-scale ship plans on the floor is a common practice in modern shipbuilding.
40. *Haifangdang*, II, pp. 114–115.
41. This is recorded in Shen Baozhen's letter to the *Zongli yamen* in October 1869. Shen explained that this production procedure took time, and also complained that the foreign technicians needed to be urged to show the Chinese workers those skills *Haifangdang*, II, p. 193.
42. *Haifangdang*, II, p. 160, p. 183.
43. *Haifangdang*, II, p. 52; *Chuanzheng zouyi huibian* 19:20ab. Lin Qingyaun gives us the technical specification of the machinery, a horizontal engine that ran at 80 revolutions per minute. It had two boilers and four furnaces. The boiler produced 2.75 times the atmospheric pressure. The boat sailed at a speed of ten knots. Xin Yuanou also gave the same data. Lin Qingyuan, *Fuzhou chuanzhenju shigao*, (revised edition), p. 134; Xin Yuanou, *Zhongguo jindai chuanbo gongye shi*, p. 131
44. *Haifangdang*, II, p. 229.
45. Shen Baozhen's father died in October 1870. Hence, he was on leave for mourning. *Haifangdang*, II, p. 304.
46. Figures from David Pong, *Shen Pao-chen*, p. 219.
47. The school was temporarily located in southern Fuzhou when the Navy Yard's construction work was carried out. The School gradually took shape when the foreign teachers arrived and the navy yard's construction work progressed, and was moved to the site of the Navy Yard in early July 1867. *Haifangdang* II, p. 59.
48. The term "front" and "rear" only indicated their location in the navy yard. And term qian xuetang and hou xuetang are first seen in Shen Baozhen's December 1873 memorial, however, the Division of English

and French school had started before May 1867. *Haifangdang*,II, p. 472; Lin Qingyuan, *Fujian chuanzhenju* (revised edition), p. 123.
49. Steven A. Leibo, *Transferring Technology to China*, p. 116.
50. Prosper Giquel, *The Foochow Arsenal*, p. 32.
51. Prosper Giquel, *The Foochow Arsenal*, p. 18.
52. Prosper Giquel, *The Foochow Arsenal*, p. 18.
53. M.M. L. Rousset taught physics and chemistry and L. Medard mathematics. Prosper Giquel, *The Foochow Arsenal*, p. 17.
54. Prosper Giquel, *The Foochow Arsenal*, p. 18.
55. Steven A. Leibo, *Transferring Technology to China*, p. 114.
56. Steven A. Leibo, *Transferring Technology to China*, p. 115.
57. Prosper Giquel, *The Foochow Arsenal*, p. 21.
58. Prosper Giquel, *The Foochow Arsenal*, p. 20.
59. Prosper Giquel, *The Foochow Arsenal*, p. 21.
60. Prosper Giquel, *The Foochow Arsenal*, p. 22.
61. Xiong Yuezhi, *Xixue dongjian*, pp. 266–277.
62. The subjects also included geology, mineralogy, international law, and political economy. Knight Biggerstaff, *The Earliest Modern Government Schools in China*, p. 123, 127–128.
63. For more details, see Knight Biggerstaff, *The Earliest Modern Government Schools in China*, pp. 154–199; Xiong Yuezhi, *Xixue dongjian*, pp. 266–277.
64. By the 1870s, Rong Hong tried to persuade Zeng Guofan to set up an engineering school. In October 1868, when he was reporting to the imperial court the result of shipbuilding and gun-making at the Jiangnan Arsenal, Zeng Guofan mentioned that he wished to establish an engineering school and select bright pupils to learn Western technology. He expressed that in the future the Chinese people would understand the technology without instructions from foreign experts. Yung Wing, *My Life in China and America* (New York, 1978, reprint of 1909 edition), pp. 168–169; *Yangwu yundong*, vol. 4, p. 18.
65. Prosper Giquel, *The Foochow Arsenal*, pp. 19–22.
66. *Yangwu yundong*, vol. 5, pp. 105–106.
67. *Haifangdang*, II, pp. 325–326, 332, 346–350, 363–372, 385–386.
68. *Yangwu yundong*, vol. 5, p. 82. In December 1873, Shen Baozhen memorialized that he was ready to dismiss the foreign technicians and requested permission to send students abroad to carry on with shipbuilding. *Yangwu yundong*, vol. 5. p. 140.
69. The first iron vessel was the *Vulcan*, a passenger barge built in 1819 on the Clyde Canal. The first sea-going iron vessel was the *Aaron Manby*, built on the Thames in 1822. The world's first iron-hulled warship was the English East India Company's *Nemesis*, which led the British expedi-

tion force in the first Opium War. The success of the *Nemesis* did not kick off a rush into warship building with iron, although iron auxiliaries were employed by British, French, and the United States navies. For more information about the history of iron ship construction in the nineteenth century, see Robert Gardiner (ed.), *Steam, Steel and Shellfire: The Steam Warship 1815–1905* (London: Conway Maritime Press, 1992).

70. For more details on hull construction, see A. M. Robb, "Ship-building," in Charles Singer et al. (eds.), *A History of Technology*, vol. 5, (Oxford: Clarendon Press, 1958), pp. 350–390.
71. The first successful application of the compound engine to a sea-going ship was made in 1854. In the 1880s, engineers developed boilers that could generate steam pressure higher than 100 lbs., and could be expanded in three stages. For more details about marine compound engines, see Edgar Smith, *A Short History of Naval and Marine Engineering* (Cambridge: Cambridge University Press, 1938), pp. 174–86; Richard Sennett, *Marine Steam Engine* (London: Longmans, Green and Co., 1885), pp. 10–12.
72. FO 233/85, J.G. Dunn, "Description of the Foochow Arsenal," 11th September 1873.
73. *Yangwu yundong*, vol. 5, p. 212.
74. *Haifangdang*, II, pp. 472–473; *Yangwu yundong*, vol. 5, p. 140.
75. *Yangwu yundong*, vol. 1, pp. 26–78.
76. In August 1874, he memorialized again stressing that the method of building ironclads could be learned by upgrading hull construction and the steam engine. *Yangwu yundong*, vol. 5, pp. 148–149.
77. Shen sent Giquel to Europe to purchase the compound engine from Maudslay's, a leading marine engineering firm in Britain, and a set of iron frames from France in March. Giquel was also to hire technicians to teach the new technology. *Yangwu yundong*, vol. 5, pp. 149–150.
78. In September, Ding Richang, former manager-general of the Jiangnan Arsenal, former governor of Jiangsu, and the then deputy of Li Hongzhang, was appointed as Shen Baozhen's successor. In December, he was also appointed governor of Fujian. Ding Richang resigned from his post in the Fuzhou Navy Yard in April 1876. Wu Zancheng, one of Li Hongzhang's protégés and governor of the metropolitan area, was appointed to direct the Fuzhou Navy Yard in April 1876. Yet, he stayed on until October 1879, and Li Zhaotang, another of Li Hongzhang's protégés, formerly general manager of the Tianjin Arsenal and then chief minister of the court of imperial entertainments, was appointed the imperial commissioner of the navy yard.
79. *Haifangdang*, II, p. 697.
80. *Haifangdang*, II, p. 691; *Yangwu yundong*, vol. 5, p. 182.

81. Although the manufacturer in England refused to sell the machinery for making the brass tubes, the foreign technician modified the Navy Yard's machinery to produce brass tubes and brought in new iron rolling mills. *Haifangdang*, II, pp. 701, 770–771.
82. While the technology for the compound engine and the iron frames were being learned, two Fuzhou-built 150-horsepower steam engines were to be modified under the foreign technicians' instruction. After the modification, the engines would reduce coal consumption by one third. These modified engines were fitted in the twenty-first and twenty-second steamers. *Haifangdang*, II, p. 691.
83. *Yangwu yundong*, vol. 5, p. 164.
84. *Yangwu yundong*, vol. 5, pp. 186–187.
85. *Yangwu yundong*, vol. 5, p. 199.
86. For more details of the training of the navigation students, see Knight Biggerstaff, *The Earliest Modern Government Schools in China*, pp. 233–235.
87. *Yangwu yundong*, vol. 5, p. 207.
88. *Yangwu yundong*, vol. 5, p. 207.
89. *Yangwu yundong*, vol. 5, p. 191. Ding Richang had ever suggested to the *Zongli yamen* that the Fuzhou Navy Yard should establish a direct link with foreign shipyards for obtaining latest technology. *Haifangdang*, II, pp. 626–627.
90. *Yangwu yundong*, vol. 5, p. 191.
91. *Yangwu yundong*, vol. 5, pp. 233–236.
92. A report of their return after three years of study appears in the *Chuanzheng zouyi huibian* dated 10th May 1886. Knight Biggerstaff, *The Earliest Modern Government Schools in China*, p. 236.
93. In 1876 Giquel went to Tianjin to meet Li Hongzhang and to discuss the matter of sending students abroad. In the meeting, he suggested to Li that the Fuzhou-built ships were old-fashioned and the navy yard should build cruisers and torpedo launchers. Li agreed with him and wrote to Wu Zancheng, the newly appointed director of the Fuzhou Navy Yard, suggesting that the Navy Yard should build four cruisers before any ironclad was purchased. However, the navy yard did not carry out the project because of lack of funds. In December 1879, the imperial court ordered the navy yard to carry out the project if it had the technical ability to do so. *Yangwu yundong*, vol. 5, p. 242.
94. *Yangwu yundong*, vol. 5, p. 242.
95. The graduates modified the original design purchase in 1877. *Haifangdang*, II, p. 937.
96. *Haifangdang*, II, pp. 936; *Yangwu yundong*, vol. 5, p. 295.

97. In November 1882, Zuo Zongtang memorialized to build four more cruisers. But for unknown reasons, the Fuzhou Navy Yard only built two more for Zuo. Li Hongzhang also ordered two cruisers for his *Beiyang* fleet, which had been gradually formed under his command since 1875. Yet, the navy yard did not carry out the plan. *Zuo Wenxiang gon quanji*, Memorials, 59:52ab; *Haifangdang*, II., p. 938.
98. *Yangwu yundong*, vol. 5, p. 315.
99. *Yangwu yundong*, vol. 5, pp. 324–325.
100. Robert Gardiner (ed.), *Steam, Steel and Shellfire*, p. 111, 176.
101. A.M Robb, "Ship-building," Charles Singer et al.(eds.) A History of Technology, vol. 5, pp. 373–374; Sidney Pollard, *The British Shipbuilding Industry*, 1870–1914, p. 14.
102. For more details about the warships built with steel in the West, see Robert Gardiner (ed.), *Steam, Steel and Shellfire*, pp. 95–110.
103. *Yangwu yundong*, vol. 5, p. 253.
104. *Yangwu yundong*, vol. 5, p. 336.
105. The engineers had come back from Germany by January 1884. *Yangwu yundong*, vol. 5, pp. 336–337.
106. Another technology that attracted the Fuzhou Navy Yard after the Sino-French War was torpedoes. In the Battle of the River Min, the French fleet's torpedoes were among other French naval armament that dealt a serious blow to the Fuzhou Navy Yard's wooden-hulled gunboats. By the 1860s the term torpedo referred to various types of naval mines ranging from floating mines to those explosive devices that were carried by submarines or swimmers. The self-propelled missile-shaped torpedo we know today was invented in 1864. Its propeller was driven by the compressed-air engine, which used compressed air to push the piston. After the Sino-French War, the Fuzhou Navy Yard ordered a torpedo boat from Germany and ten torpedoes from France in order to imitate such technology. The new boat and torpedoes arrived at Fuzhou in 1886. A French Division student who learned torpedo technology in Germany was in charge of the torpedo production. A torpedo workshop was built and machinery was ordered from Germany in the same year. The acquisition of the torpedo-producing facilities enabled the navy yard to equip its new ships with torpedoes. Unfortunately, the available documents reveal little about how much the navy yard could do in terms of producing the torpedo parts itself. *Yangwu yundong*, vol. 5, p. 351.
107. *Yangwu yundong*, vol. 5, pp. 311–312.
108. It is unclear when the Fuzhou Navy Yard started steel-making. From a memorial of January 1886 by the director of the navy yard, it is evident the navy yard had made steel masts for the second cruiser. The director reported that steel sheet was easy to make but steel-making could relieve

the difficulty of obtaining wooden material. *Yangwu yundong*, vol. 5, p. 325.
109. *Yangwu yundong*, vol. 5, p. 346, 354.
110. The British naval commander, Admiral Lang, who was engaged by Li Hongzhang to train the Beiyang fleet's officers, told the British vice-consul at Fuzhou in private that the ship was a complete failure. According to the British Consul at Fuzhou's report, the ship was built according to Li Hongzhang's requirement. Li hoped the ship could sail into the River Beihe as far as Tianjin. FO 228/876, "Report on the Foreign Trade of the Port of Foochow for the Year 1888," 9th March 1889; FO 228/889, Hurst to Sinclair, 21st May 1890.
111. Zhang Zhidong took over an improved composite ship and named it as *Guangjia*. It was launched in August 1887 and was fitted with a newer triple expansion 1600-horsepower engine purchased from Britain. *Yangwu yundong*, vol. 5, p. 365, pp. 368–369, 377.
112. *Yangwu yundong*, vol. 5, p. 388. The Fuzhou Navy Yard reported to the court that that the frame was iron. It is likely that using iron rather than steel to build the frame was technically easier and financially cheaper.
113. A detailed list of the upgraded machinery can be found in Li Qingyuan, *Fuzhou chuanzhenju*, pp. 197–200.
114. *Yangwu yundong*, vol. 5, p. 217.
115. *Yangwu yundong*, vol. 5, p. 223.
116. *Yangwu yundong*, vol. 5, p. 256.
117. *Yangwu yundong*, vol. 5, p. 267.
118. *Yangwu yundong*, vol. 5, p. 377.
119. *Yangwu yundong*, vol. 5, p. 297, 375–376.
120. In 1883, the director of the Fuzhou Navy Yard proposed to build a dry dock but failed to obtain the court's permission. In January 1886, the navy yard memorialized for permission to carry out this project again, but Prince Chun, the chief minister of the *Haijun yamen* rebuffed the proposal on the grounds of government financial strain. *Haifangdang*, II, p. 938; *Yangwu yundong*, vol. 5, pp. 321–323, 340, 382–384, 396, 407, 415–416.
121. *Haifangdang*, II, p. 771.
122. Jouvet left for France in September 1877, although three technicians were retained for one more year until September 1878. *Yangwu yundong*, vol. 5, p. 212, 274.
123. The earliest document I have seen that mentioned the term of *gongcheng chu* is a memorial of October 1883. *Yangwu yundong*, vol. 5, p. 295.
124. *Yangwu yundong*, vol. 5, p. 237.
125. *Yangwu yundong*, vol. 5, p. 187, 236, pp. 251–254.
126. *Yangwu yundong*, vol. 5, pp. 263–265.

127. *Li Wenzhonggong quanji*, Memorials, 55:14a–15b.
128. *Chuanzheng zouyi huibian*, 41:8a–11b; Knight Biggerstaff, The Earliest Modern Government Schools in China, p. 239.
129. It is noteworthy that some of the returned students from the Chinese Educational Mission to the United States, initiated by Zeng Guofan in 1872 and withdrawn in 1882, were assigned to the Schools of the Fuzhou Navy Yard to learn navigation. Knight Biggerstaff, *The Earliest Modern Government School in China*, p. 226.
130. *Chuangzheng zhouyi*, 24:12a–14b.
131. L. Médard, who had been teaching in the school since 1868, was retained. Another Frenchman whose Chinese name was Deshang was recruited, although it is not clear when he was employed. The Sino-French War forced Médard to resign and an English teacher, F. T. Richards was recruited in August 1886. *Chuanzheng zouyi huibian*, 15:32a, 28:9ab; Knight Biggerstaff, *The Early Modern Government Schools*, p. 224.
132. *Yangwu yundong*, vol. 5, p. 209.
133. The Dreadnaught was the benchmark of the early twentieth-century warship technology. It had ten 12-inch guns housed in five turrets. It was equipped with the state-of-art 22,500 horsepower steam turbine that could reach the top speed of 21 knots. It had a belt-armor as thick as 11 inches. It was 18,000 tons and 527 feet long, with an 82 foot beam, and a 26 foot draft.
134. *Chuanzhen zouyi huibian*, 45: 14ab.
135. The Nanyang despatched the Fuzhou-built *Kaiji* and its two sister cruisers to assist the *Beiyang* fleet. Guangdong sent the Fuzhou-built *Guangjia*, *Guangyi*, and *Guangbing*. Shen Chuanjin, *Fuzhou chuanzheng ju*, p. 261.
136. Liu Haifeng and Zhuang Minshui, *Fujian jiaoyu shi* (Fuzhou: Fujian jiaoyu chubanshe, 1996), pp. 247–249.
137. Shen Chuanjin, *Fuzhou chuanzheng ju*, pp. 335–336; Lin Qingyuan, *Fujian chuanzhenju shigao* (revised edition), p. 458.

CHAPTER 6

To Buy Or to Build?

Chapter 5 has shown that the technical progress of the Fuzhou Navy Yard slowed down after the late-1880s. A major reason for this was the shortage of funds. Since the mid-1870s, the navy yard had to handle the mounting costs of running the retained steamboats and building new ships with a fix budget. Reduced support limited the scale of facilities upgrade and affected the progress of shipbuilding works. The problem had reached such a point between 1889 and 1894 that it could not launch any shipbuilding project. The problem had much to do with a flawed funding arrangement in a fiscal system that did not see the financial needs for long-term investment in technology. This chapter argues that the weakness of the Qing government's fiscal system had made it unfit to fund large-scale industrial operations and the government's shift in policy had exposed such weaknesses that technology transfer was crippled.

THE ORIGINAL PLAN

Before Zuo Zongtang submitted the proposal for establishing the Fuzhou Navy Yard in 1866, he took care in planning how to fund it and how the steamers could be employed. As to the running costs, Zuo estimated that the navy yard would need 40,000 taels monthly. He earned the governor of Guangdong and the governor of Zhejiang's agreement to pool financial resources together to support the project. Fujian would put in 20,000 taels monthly and Zhejiang and Guangdong each committed 10,000 taels monthly.[1]

© The Author(s) 2022
H.-ch'un Wang, *Western Technology and China's Industrial Development*, https://doi.org/10.1057/978-1-137-59813-4_6

As to the use of the steamboats, Zuo Zongtang planned that coastal provinces could employed them for coastal patrol against pirates. To pay for the running costs, these vessels could carry merchandise and offer protection to merchants. They could also be leased to merchants for shipping tributary grains. He suggested that foreign instructors should be hired to train the seamen, who could be recruited from the area of Ningbo, Zhejiang.[2] However, despite careful planning, soon after the imperial court permitted the project, Zuo was appointed governor-general of Shaanxi and Gansu. This unexpected appointment was to have an impact on how the Fuzhou Navy Yard would be run and funded. The first question was who would run the navy yard.

The imperial edict ordered Zuo's successor in Fujian to take over the navy yard project.[3] Yet, the development in Fuzhou went beyond the imperial court's expectation. In October 1866, Fujian senior officials forwarded to the throne a petition presented by the local gentry led by Shen Baozhen, former governor of Jiangxi who was at the time mourning his mother's death in his hometown of Fuzhou, asking for the postponement of Zuo Zongtang's departure.[4] The petitioners argued that Zuo Zongtang was the only person who could direct Giquel and d'Aiguebelle to initiate such an important project. In the same month, Zuo Zongtang memorialized requesting permission to delay his departure for the northwest on the grounds of the need to finalize the contract with the two French officers and prepare for the building work. In the same memorial, he recommended that Shen Baozhen be appointed the director of the Fuzhou Navy Yard.[5] The imperial court permitted Zuo's late departure and ordered Shen Baozhen to direct the navy yard project. It also ordered senior Fujian officials to manage the funding of the project.[6] Zuo obviously had wanted to keep his influence at the Fuzhou Navy Yard and have it run by people he trusted. Shen Baozhen had formerly been a close-associate of Zeng Guofan, and, by 1867, had started to associate himself with Zuo Zongtang.[7] By successfully recommending Shen Baozhen to the imperial court, Zuo Zongtang out-maneuvered the succeeding governor-general of Fujian and Zhejiang away from the navy yard project.[8]

Nevertheless, Zuo Zongtang did not win full control of the navy yard. The imperial court instructed that all memorials that form the Fuzhou Navy Yard should bear both the seals of Zuo Zongtang and three senior Fujian officials (governor of Fujian, governor-general of Fujian and Zhejiang, and Fuzhou Manchu General, who managed the revenue of the

Fujian maritime customs.). This arrangement limited Shen Baozhen's authority, especially in raising funds.

Zuo Zongtang's removal from Fujian was to change the navy yard's funding arrangements. Zuo revised his plan to fit his new position, and hence diverted the 20,000 taels monthly from Zhejiang and Guangdong to his northwest campaign with the result that the navy yard had to be funded totally from Fujian. He estimated that the navy yard would need over 600,000 taels per year to run, and building the navy yard and purchasing machinery would need an additional 430,000 taels.[9] Hence, he budgeted that 400,000 taels should come from the 40 percent of the Fujian maritime customs revenue of that year and the income from the Fujian transit dues would supplement any shortfall. This revision earned the imperial court's agreement.

As the historian David Pong has pointed out, this budget had left out the potential costs of vessel maintenance.[10] Such a miscalculation was to cause financial difficulties soon after the steamers were completed and up and running. Moreover, the directors of the navy yard had no influence on the Fujian authorities and the progress (or failure) was not directly linked to provincial officials' achievement. The Fujian authorities' willingness to maneuver financial resources to support the navy yard would, therefore, depend on the imperial court's attitude. Such a separation of management from finance was to make it difficult for the Fuzhou Navy Yard to obtain funds. This was a sharp difference with the Jiangnan Arsenal's funding arrangement.

Although the Jiangnan Arsenal's funding scheme took some time to shape up, its funding and managerial authorities lay in the same hands.[11] Li Hongzhang established the arsenal in 1865, and in 1867, his patron, Zeng Guofan, then governor-general of Jiangsu, Jiangxi, and Anhui, took direct control of the arsenal and obtained imperial approval for allocating four percent of the Shanghai maritime customs revenue to support the Jiangnan Arsenal.[12] In March 1869, Zeng's successor in Nanjing obtained the court's permission to transfer another four percent of Shanghai's customs revenue to fund the arsenal.[13] The arsenal had built 11 gunboats by 1875, nine of them being employed and paid by the Jiangsu authorities.

The crucial part of this arrangement was that the general manager (*zongban*) of the arsenal was one and the same as the Shanghai circuit intendant (*daotai*), who managed the revenues of the Shanghai maritime customs. Moreover, until 1895, the position was always occupied by one of Li Hongzhang protégés, although Li's influence might have been diminished due to his long tenure in Zhili.[14] Furthermore, Li Hongzhang held the position of one of the four supervisory directors (*duban*).[15] This arrangement ensured that between 1867 and 1895 550,000 to 600,000 taels from the Shanghai maritime customs and other sources flowed into the arsenal's coffer annually.[16]

The Fuzhou Navy Yard's arrangement was different. Neither Zuo Zongtang nor Shen Baozhen exerted a strong influence on the Fujian authorities that managed the Fujian maritime customs and the transit dues. Shen Baozhen could not submit a memorial without its bearing Fujian senior officials' seals. That is to say, only the imperial court could make the Fujian authorities deliver funds to the navy yard. In the 1860s, the imperial court's eagerness to engage in self-strengthening and Shen's prestige made up for any flaw in the arrangement. Shen Baozhen was to lead the navy yard through the early financial troubles that derived from the costal provinces' reluctance to take over Fuzhou steam vessels.

The Immediate Problem

Zuo Zongtang had assumed that the coastal provinces would take over the Fuzhou steamers for coastal patrol and the running costs would be paid by shipping tributary rice.[17] Hence he planned to have Fuzhou steamboats built as "half-military-half-merchant" (*banbing banshang*) vessels.[18] That is to say, they had cargo bays as well as guns and magazines. They were more like armed transport. Ironically, it was such versatility that affected the effectiveness of the steamboats' in shipping tributary rice.

Since the 1850s, due to the heavy silting on the Grand Canal, Zhejiang and Jiangsu officials had hired merchant junks for shipping tributary rice via the sea route to Tianjin.[19] The junk fleets would be escorted by government war junks or, occasionally, by hired foreign-owned gunboats.[20] Junks were allowed to carry cargo back to the south in order to make the journeys profitable. Besides, since 1867, Chinese merchants and senior officials had been discussing the possibility of forming a joint-stock steam navigation company that included shipping tributary rice as part of its business.[21] Zuo's plan might have fitted into such an environment.

However, the Fuzhou steam gunboats' cargo space and the firepower would compromise each other, although they might have been effective in fighting less-well-armed pirates. Senior officials knew the problem very well. In January 1872, Li Hongzhang told Zeng Guofan that the Fuzhou-built steamers loaded less cargo and sailed more slowly than merchant steamers.²² Therefore, the Fuzhou steamers were unattractive to merchants and senior officials who would like to ship tributary rice as efficiently as possible, especially when the Fuzhou-built steamers' running costs were high.

For example, the 150-horsepower Fuzhou boat *Wannian Qing*, which had the largest cargo capacity among the early Fuzhou steamers, could carry only 600 tons of cargo. It was manned by 100 seamen, whose salaries were 2100 taels in total a month.²³ It probably consumed 200 taels of coal a month.²⁴ However, employing traditional junks would not need such high costs. The government would only spend around 1200 taels on hiring a large junk that had the capacity of 3000 *shi* (around 150 tons) for one trip of tributary rice shipping.²⁵ Therefore, coastal provinces would find it hard to employ Fuzhou steamboats on a permanent basis and pay the costs by shipping tributary rice.

Before the costal provinces took over the steamboats, the Fuzhou Navy Yard had to pay their running costs, which were not allowed for in Zuo Zontgang's funding arrangement. Shen Baozhen was quick to address the problem. He obtained extra funding first. In 1869 he persuaded the Fujian authorities to obtain the court's permission to appropriate funds from the opium tax levied at Fuzhou and Xiamen.²⁶ This source gave approximately 78,125 taels to maintain the ships.²⁷

Yet, when the navy yard built more steamboats the coastal provinces were still slow to take them over. In October 1870, after four boats were completed, only Zhejiang province took over one boat probably because of the Zhejiang-Fujian connection.²⁸ Hence, the money from the opium tax was not enough to maintain the newly completed steamers. In September 1871, the Fujan senior officials suggested to the imperial court that either the coastal provinces should take over the Fuzhou vessels or additional funds had to be found.²⁹ Hence the *Zongli yamen* wrote to the senior officials of coastal provinces, urging them to take over the Fuzhou steamers.³⁰

After months of discussion among senior officials, in 1872, Zhili, Shandong, Zhejiang, and Guangdong took over one Fuzhou steamers respectively.³¹ Giving away four boats helped to ease up some financial pressure but did not solve the problem. The navy yard still had four steamers in hand and the maintenance funds were on the verge of being depleted. Hence, the navy yard had to pay out the money from its shipbuilding budget.³² For a better solution, Shen Baozhen and senior officials had to come back to the question of making Fuzhou vessels effective.

As mentioned above, the already-built Fuzhou vessels were not considered to be effective as cargo ships. In the political atmosphere of the self-strengthening, senior official of coastal provinces might have taken over a couple of the vessels as a gesture of showing their support to the Fuzhou shipbuilding project. Hence, occasionally Li Hongzhang employed Fuzhou vessels to ship tributary rice to Tianjin. Yet, Zhejiang, Shandong, Guangdong, and Li took the Fuzhou gunboats mainly for coastal patrolling. They would not have wanted more than a couple of them.[33] Therefore senior officials considered that the Fuzhou Navy Yard should build vessels that could attract merchants. Li Hongzhang made such a suggestion in June 1872.[34] The suggestions fitted into the plan Li and his associates had made for founding the China Merchants' Steam Navigation Company. Hence, by February 1873, Shen Baozhen ordered Giquel to enlarge the cargo bay of the twelfth steamer, which was being built, and made the thirteenth, fourteenth, and fifteenth steamers follow suit.[35] The new steam navigation company seems to have been interested and, by February 1874, showed willingness to take over the three modified vessels.[36] Yet, by 1874, it had only taken two despite the fact that Shen Baozhen promised the company that it did not have to pay any rent or insurance.[37]

The plan to build commercial vessels did not work well. Hence, extra funds had to be found to relieve the problem. In December 1872, Zuo Zongtang obtained the court's permission to divert 20,000 taels monthly from the 50,000 taels that Fujian was supposed to provide for him in support of his northwestern campaign.[38] This money was paid in installments for 13 lunar months beginning in January 1873. This gave the navy yard 260,000 taels in total.[39] Besides, the creation of the Taiwan Defense Fund in response to the Taiwan Crisis of 1874 threw a life-line to the Fuzhou Navy Yard. In January 1875, Shen Baozhen obtained the court's permission for the Taiwan local authorities to pay the running costs of the Fuzhou vessels on the grounds that they were employed for the defense of Taiwan.[40] Hence, the Taiwan fund paid the soaring costs of boat maintenance, which had reached nearly 480,000 taels cumulatively in 1874.[41]

Therefore, by obtaining extra funds, the Fuzhou Navy Yard managed to pay the boat maintenance costs. In March 1875, Shen Baozhen produced his fist financial report as the director of the navy yard. Between December 1866 and August 1874, the navy yard had completed fifteen steamboats. Up to date, four of them were serving coastal provinces and one served the China Merchants' Steam Navigation Company. The navy yard had received 8,960,000 taels for shipbuilding and navy yard construction. It had a small balance of 3640 taels. Shen created a steamboat maintenance account, which received 429,372 taels from the opium tax and 192,458 taels from the shipbuilding funds.[42]

In short, Zuo Zongtang's preliminary arrangement of giving away Fuzhou steamers did not work out. The half-military-half-merchant features of the Fuzhou steamers had not created any incentive for taking them over either by coastal provinces nor merchants. Hence, the increasing costs of maintaining the steamers became a financial burden to the navy yard. Yet, Shen Baozhen, senior Fujian officials, and the *Zongli yamen* worked together to obtain extra funds and, by the end of 1873, persuaded provinces to take over five steamboats. In additions, the establishment of the Taiwan Defense Fund and the opium tax provided financial resources for retaining the vessels in the Fuzhou Navy Yard.

The difficulty of giving away Fuzhou steamboats shows that the navy yard as an infant industry was suffering from having to prove its steamers' effectiveness before they could become effective. The steamers were not attractive either as warships or merchant vessels, even if they were free.

However, the problem lay not only in an unattractive product but also in a society that had not been able to accept a new technology that demanded social and technological changes. In the 1870s, Qing navies had little capacity in accepting steam warships. Both funds and naval officers were in short supply. Without an overhaul of the naval forces, coastal provinces would only want to employ no more than a couple of gunboats for costal patrol. Unfortunately, Li Hongzhang's suggestion for disbanding traditional navies and decommissioning war junks in 1872 attracted no attention probably because senior officials did not see the need for such drastic measures.[43]

The market for merchant vessels was small as well. In the 1870s, the only potential buyer of Fuzhou vessels was the China Merchants Steam Navigation Company, which had the monopoly of shipping tributary rice. Senior officials estimated that the company needed between 10 and 20 vessels to carry out its business.[44] Yet, the company purchased foreign-built vessels and took over a few Fuzhou vessels in the 1870s.

Thirdly, the supply of fuel was another concern for coastal provinces. China's fuel market relied heavily on imported coal. China had not opened any substantial coal mines with Western machinery until Shen Baozhen initiated the Jilong Coal Mine in Taiwan in 1875 to supply the Fuzhou Navy Yard. Li Hongzhang only opened the Kaiping Coal Mine in 1880. Hence, domestic coal was in short supply through the 1870s.[45]

Therefore, despite senior officials' efforts in 1875, the Fuzhou Navy Yard still retained its ten steamboats. Nor did it intend to carry on building merchant vessels because Shen Baozhen and senior officials did not consider it an option.[46] Although the financial difficulty of 1871–1874 had been eased, the long-term problem of the Fuzhou Navy Yard was to be exposed after Shen Baozhen was removed from Fuzhou to be governor-general of Jiangsu, Jiangxi, and Anhui in 1875.

The Long-term Problem

From the mid-1870s, the Fujian authorities started to reduce the funds that were delivered to the Fuzhou navy yard, and by early 1875, their arrears were more than 200,000 taels. For the next ten years, the Fujian authorities never sent the 50,000 taels monthly budget to the navy yard in total. In consequence, when the Guangdong support stopped in 1889, it took four years for the navy yard to complete a cruiser. Successive directors kept memorializing in vain, requesting the imperial court to instruct the Fujian authorities to deliver the money.

What prevented the Fujian authorities from delivering the money was not any ideological opposition against Western technology. The imperial court had set up the Maritime Defense Fund, spending millions of taels from the 1870s to purchase warships from abroad. It permitted Guangdong's purchase of a foreign-owned dockyard in Huangpu in 1876 and Li Hongzhang's establishment of the Dagu Dockyard in Zhili province in 1880. The former built gunboats and the latter was aimed at servicing Li Hongzhang's purchased warships. Nor did Fujian senior officials have anything against the Fuzhou Navy Yard or its directors. The problem had much more to do with the inability of the Qing fiscal system to support a modern navy yard and the director's inability to manage the navy yard amidst the changing tide of opinion on warship building. The discussion should start from an overview of the fiscal system and revenues and expenditures between the 1870s and the 1890s.

The Qing political system was highly centralized. The imperial court held absolute power in the appointment of officials, and in assigning missions to military commanders, even though it needed to take advice from the Grand Council, the six Boards, and provincial senior officials to formulate policies. In terms of the fiscal system, the Board of Revenue, which managed the national revenue and expenditure, could not issue instructions to governors and governor-generals. On the provincial level, administration commissioners, grain intendants, salt intendants, and customs intendants, who collected and managed taxes, were directly appointed by the emperor and, before the Taiping Rebellion, were not directly subordinate to governors or governor-generals.

The tax revenues collected by provincial authorities were divided into three parts. One part, which consisted of the central government funds (*jingxiang*), was remitted to the Board of Revenue. The second part, which was withheld taxes (*cunliu*), was retained to pay for the expenditures of provincial governments. The third part, which might be translated as subsidiary funds (*xiexiang*), was remitted to various destinations for funding specific public works, military operations, famine relief, or for supporting provinces whose revenues were unable to meet expenditures. In the case of the maritime customs revenue, 40 percent of it was the central government fund and 60 percent was meant to be the withheld taxes and subsidiary funds.[47]

The overall structure of the Qing government budget was created in the Kangxi reign (1662–1722). Although the Board of Revenue was well aware of the changes in tax revenues and expenditures, it did not try to change substantially the tax quota and the budget, but tended to take temporary measures to deal with the discrepancies between expenditure and income. In the early Qing dynasty, the imperial court saw curbing government spending as ethical but raising tax rates as morally and politically unacceptable.[48] Yet, the normal budget did not consider unexpected expenditures, such as wars, river conservation, or famine relief. To support these unbudgeted spending, the government avoided raising the land and poll tax, which was the principal tax (*zhengshui*), but tended to use temporary measures such as raising the rates or adding new items to the miscellaneous taxes (*zashui*), or selling limited numbers of imperial degrees and government offices. In this way, the early Qing government managed to finance major military campaigns.[49]

However, the central government had been gradually losing control over the taxation mechanism since the 1850s because of the protracted Taiping Rebellion, which stretched government finance to its limit. Not only were the military expenditures enormous, the civil war also disrupted tax collection, damaged the economy, and destroyed the land registry of the Lower Yangzi area, the largest source of the land and the poll tax.[50] Both central and provincial treasuries were depleted. New local taxes were imposed, and the rate of miscellaneous taxes was increased. In addition, the new Chinese maritime customs, which was formed in 1854 and managed by Westerners, had become a stable source of income. In the case of emergencies, provincial governments obtained short-term loans from Chinese or foreign merchants or banks.

The raising of new local taxes since the 1850s marked the beginning of the change of the Qing fiscal system, because new taxes were imposed to support the militia forces organized by provincial officials. Among the new taxes, the transit dues (*lijin*), instituted in 1854, was the most important one. Originally the imperial court took it as a temporary measure and wished to abolish it as soon as peace was restored. Yet, it became important government revenue and even outlived the dynasty.[51] The statutory quota of land and poll tax was around 45 million taels.[52] Since 1885, the actual income of the land tax had fallen to around 32 million taels, but the transit dues and the maritime customs had reached 30 to 38 million taels annually. Since new taxes were primarily retained by the provincial government, senior provincial officials were gaining more financial and military independence at the expense of the central government.[53]

The end of the Taiping Rebellion in 1864 did not ease the fiscal strain because of the continuous rebellions of the Nien in the north and the Moslem rebellion in the northwest. Unfortunately, there is neither data nor estimates of the expenditures of the 1860s and 1870s. However, those of the 1880s may give us an impression of the scale of military costs. According to the Accounting Records of the Reign of Guangxu (*Guangxu kuaiji lu*), the expenditure on militia troops was between 18,260,000 and 27,610,000 taels between 1885 and 1894.[54] The officials of the Board of Revenue made an even higher estimation in 1885: the costs of the Green Standard were estimated at 14 to 15 million taels a year, the militia troops at 34 million taels, and the Banner troops at more than one million taels. But the revenue of the government was just 70 to 80 million taels between 1885 and 1894.[55] That is to say, around 62.5 percent of the government revenue was spent on paying the expanded troops.

Among the military expenses, Zuo Zongtang's northwest campaign was one of the largest. It relied on subsidiary funds from other provinces, including Fujian's maritime customs. The importance of the campaign made the imperial court give Zuo full financial support. Between 1875 and 1881, Zuo spent 52.3 million taels in total, which was on average nearly 7.5 million taels a year.[56] It was 11 percent of the revenue, if we use the 1885 revenue (18,260,000 taels) as a point of comparison. Even after the campaign was over, the Qing government still had to spend roughly four million taels annually after 1884 for the up-keep of the troops stationed there.⁵⁷ Such expenditures competed directly with the Fuzhou Navy Yard, the budget of which relied on the Fujian maritime customs and the transit dues.

Compared to the spending discussed above, the Fuzhou Navy Yard's 600,000-tael annual budget does not raise eyebrows. It was only less than one percent of the tax revenue, if we use the 1885 revenue as a point of comparison. The extent of financial support the imperial court was prepared to give to the government undertakings depended on the direction of policies and how well the undertakings were managed. New ways of funding could have been introduced if they were necessary or proved effective. The first seven years of the Fuzhou Navy Yard was a result of the mixture of shipbuilding as a part of the self-strengthening policy and the good management of Shen Baozhen.

When the Fuzhou Navy Yard project started in 1866, the imperial court gave it strong support because it considered that domestic steam warship building was essential to the dynasty's survival. Yet, the Fujian local officials had felt the pressure of the financial demand for simultaneously delivering other tax quotas and the new burden of supporting the navy yard project and the northwest campaign. In November of the same year, the Fuzhou Manchu General memorialized to divert some money from the 40 percent of the Fujian maritime customs revenue, which was due the central government, for the navy yard project. He argued that 60 percent of the Fujian maritime customs revenue was 240,000 taels short of meeting the tax quotas, and the revenue from the transit dues had to pay local troops and support Zuo Zongtang's northwest campaign. The imperial court agreed to his request.⁵⁸ The court's permission shows that at the time, it was willing to set aside part of the funds due itself to support such an important project. Hence, the Fujian authorities paid out the full amount for the navy yard's capital construction, monthly budget, and even the maintenance costs as discussed earlier on. However, two unexpected events were to push Shen Baozhen's political skills to the limit.

Firstly, the navy yard's spending drew criticism from the court official Song Jin. By January 1872, the Fujian authorities had diverted to the navy yard 3,150,000 taels from the Fujian maritime customs and 250,000 taels from opium taxes. The navy yard had spent more than Zuo Zongtang's estimate but only completed half of the proposed steamers.⁵⁹ Nevertheless, the imperial court's concern was not overspent budget. When Song Jin challenged the Fuzhou Navy Yard's spending, the imperial court clearly told senior officials that it would continue to support shipbuilding if domestically built steamers were effective in defense.[60] Hence, the issue of spending and the effectiveness of the vessels were brought into the spotlight. Shen Baozhen had to justify the spending by proving the effectiveness of the project and by improving the prospect of future technical development. In December 1873, Shen assured the imperial court that the navy yard had the ability to build steamers without the foreign technicians and that he would allocate their salaries for the shipbuilding work budget in the future. At the time, he had a plan for the navy yard's second stage of development in mind. However, before Shen and the imperial court could make any further decision on the Fuzhou Navy Yard's future, the Taiwan Crisis of 1874 made a much bigger impact on the policy towards shipbuilding than any concern over spending.

During the crisis, senior officials discovered that Japan purchased two ironclads, which were built by the Confederate during the American Civil War in the 1860s. They were concerned that China might not be able to defend itself if a full-scale conflict broke out. Hence they sought diplomatic negotiations with Japan.[61] Soon after the settlement, in the maritime defense debate, senior officials discussed how to improve maritime defense. Obtaining powerful coastal defense guns and ironclads was the center of the discussion.[62]

The term "ironclad" originally referred to wooden-hulled warships plated with iron armor. Yet, by the 1870s, the meaning had been corrupted to refer to armored warships of part or full iron or steel construction. The invention of the ironclad was the consequence of the impact on wooden hulls by the development of naval ordnance technology. By the nineteenth century, traditional solid round gun shots might punch through heavily-built wooden hulls, which would be reinforced by complex wooden structure, but seldom sank them. Nor could they do too much damage to the sail nor immobilize ships. Yet, the elongated explosive shell, which was introduced in the 1820s, had a devastating effect because it increased the force of penetration and the accuracy of gunnery. Besides, the explosion killed more ship crews and caused both fire and much more serious damage. Hence, since the 1830s, Western navies built ships plated with wrought iron to counteract shells.[63] Furthermore, when the steam engine was introduced into warship building in the 1840s, armoring for protecting the machinery became an important issue. Since the Crimean War (1854–1856) and the American Civil War (1861–1865), ironclads proved effective in fighting against wooden-hulled warships.

Therefore, from the mid-1860s, a race to build ironclads began among Western powers. It was the end of the wooden-hulled warships.[64]

Qing senior officials were not totally ignorant of ironclad technology. By 1874, Western ironclads had been visiting China's treaty ports. Officials boarded the warships and witnessed their strength.[65] Yet, these visits in themselves did not mean that Qing senior officials had carefully studied ironclad technology or had started to consider how China's navy yards could transfer such technology. Nor did they consult any engineers on this matter. Their decision in 1874 and 1875 was that domestic navy yards should carry on building warships, improve their technology, and eventually build ironclads.[66] They were eager to fulfil the urgent need for forming the *Beiyang* and the *Nanyang* fleets with purchased ironclads. Hence, the imperial court set up a Maritime Defense Fund with an annual budget of 4,000,000 taels for the purpose, drawing from the transit dues and half from the 40 percent of maritime customs revenue.[67]

The eagerness to obtain ironclads had helped Shen Baozhen to push through the Fuzhou Navy Yard's agenda in 1874 and 1875. The imperial court agreed to allow the navy yard to carry on building warships and send students to Europe in the hope that the navy yard would be able to build ironclads in the future. However, when Shen was laying the foundations for the future technical development of the navy yard, the Fujian authorities had started to reduce the amount of money delivered to it.

Between the ninth lunar month (October/November) of 1874 and the first month of 1875, the arrears had risen to more than 400,000 taels. Shen fully understood the reason. 60 percent of the 2,360,000 taels of the Fujian maritime customs' revenue was around 1,400,000 taels. Before delivering the Fuzhou Navy Yard's budget, this amount of money had to be used to pay for six different obligations, which amounted to 1,160,000 taels. These quotas included military funds for the northwestern campaign and contributions to the central government. The remaining 240,000 taels could not even cover the Fuzhou Navy Yard's arrears. Moreover, the Fuajian authorities still had other tax quotas to pay. The Fujian maritime customs' 60 percent revenue was 620,000 taels short.[68] Shen requested that the imperial court instruct the Board of Revenue to reduce the Fujian maritime customs' burden.[69] Yet, the Board only urged the Fujian authorities to do their best to collect taxes.[70] In late 1875, the Fujian authorities finally remitted 10,000 taels to the navy yard. It was only a small fraction of the arrears.[71]

The 620,000 taels shortage was a much more serious problem than the shortage of 240,000 taels in 1866. Even so, senior officials would not let this shortage to jeopardize the Fuzhou Navy Yard's shipbuilding project. Thanks to the Maritime Defense Fund, senior officials incorporated the Fuzhou Navy Yard into the new maritime defense policy and drew part of the navy yard's budget from the Fujian maritime customs' contribution to the Maritime Defense Fund. In December 1875, the Board of Revenue obtained the court's permission to divert 20,000 taels monthly from the 40 percent and 30,000 taels from the 60 percent. The new arrangement started from the second lunar month (February/March) of 1876.[72]

Due to the new arrangement, the 20,000 taels became the only relatively stable income for the Fuzhou Navy Yard. Between early 1876 and early 1880, the Fujiang authorities delivered the full amount from 40 percent of the Fujian maritime customs revenue. However, the arrears from the 60 percent were 33 lunar months. Between the seventh lunar month (August/September) of 1874 and the twelfth lunar month of 1879 (January 1880) (64 lunar months in total), the navy yard received 2,540,000 taels from the Fujian maritime customs, leaving 690,000 taels in arrears. The navy yard's average monthly income was around 37,352 taels.

Few senior officials went to rescue the Fuzhou Navy Yard's financial difficulty, not even Li Hongzhang, who was the greatest promoter of Western technology in the late Qing, and Shen Baozhen, who was formerly the director of the navy yard. Between 1877 and 1880, the navy yard had felt that this amount of money was not enough for carrying out the new composite cruiser project, which was budgeted at 400,000 taels.[73] The director started to seek support from Shen and Li but without success. Li and Shen had their own reasons to be reluctant to lend their political supports to the navy yard. Li never entertained the idea of developing domestic shipbuilding.[74] He knew that it took time and money for a Chinese domestic shipbuilder to develop the technology. He thought that domestically-built vessels would be expensive and less effective.[75] Hence, he suggested that purchasing ships from abroad would be more viable.[76] Nevertheless, the navy yard's training program had met his need for naval officers and onboard engineers. Hence he supported the navy yard in obtaining funds from the Fujian maritime customs and the transit dues for the training program abroad.[77] Shen Baozhen had his own consideration as the governor-general of Jiangsu, Jiangxi, and Anhui. In February 1879, Shen refused the navy yard's request for diverting funds from his account in the Maritime Defense Fund to support the cruiser project.[78] Even if Shen had more sympathy than Li Hongzhang towards the Fuzhou Navy Yard, he would have been considering his own duties in his new position and the task of building the *Nanyang* fleet, which had to spend heavily on purchasing warships from abroad out of his Maritime Defense Fund.

Furthermore, senior officials were facing a much more pressing problem, which was a potential danger of being undermined by Japan in a race of naval building up after the 1874 Taiwan Crisis. Japan was building up its modern naval force by purchasing. In 1870, Japan had decided to follow the British Royal Navy as the model of development. By 1874, the Imperial Japanese Navy had owned two steel-hulled steam cruisers, one of which was steel-armored.[79] In 1878, Japan purchased three armored frigates, one of them 37,000 tons. In 1879, Japan recruited a British naval commander to train Japanese naval cadets.

Senior Qing government officials, especially Li Hongzhang, had been paying attention to Japan's military build up. Fuzhou's 150-horsepower, 1200 to 1500-ton wooden-hulled gunboats and even 750 horsepower composite frigates of a similar size were not very attractive to them. After the Taiwan Crisis, Shen and Li had moved very fast to acquire iron Rendel gunboats (named after its designer, George Rendel) from the British shipbuilder, Armstrong, through Robert Hart. They hoped that Rendel gunboats' low costs, small displacement, but single powerful gun would offset the strength of the ironclad.[80] Furthermore, the establishment of a Chinese legation in Britain had made the information on new warship design in Britain accessible to senior Qing officials. It facilitated more purchases directly from British shipbuilders, bypassing the brokerage of Hart. Then Li Hongzhang ordered two steel-protected cruisers (1350 tons, 2600 hp, 210 × 32 × 15.5 ft), which were considered better than Japan's outdated warships.[81]

The only senior official who lent support to the Fuzhou Navy Yard was Zuo Zongtang.[82] In October 1881, he agreed to provide half of the building costs of the cruiser project. Only then could the construction work on the first cruiser, the *Kaiji*, start.[83] Also, the return of the first group of naval engineers from Europe to Fuzhou might have elevated the profile of the navy yard.[84] Their work of completing the cruiser *Kaiji*, which by the time was the most advanced domestically built steamer, impressed senior officials. Hence Zuo placed more orders. Even Li Hongzhang was tempted to do so but he later withdrew the order.[85]

Zuo's help and the Fuzhou engineers' technical ability did not change the navy yard's worsening financial difficulty. From 1880 to 1890, the Fujian authorities only remitted 3,450,000 taels, with 3,380,000 taels still in arrears. This was only half of the budget. Had the 849,919 taels from Zuo Zongtang for the composite cruiser project not been obtained, the last two composite cruisers might not have been built.

Despite Fuzhou's technical progress between 1880 and 1887, senior officials paid much more attention to purchasing warships from abroad. Li Hongzhang purchased two heavily-armored battleships, one steel-protected cruiser, and two steel-armored cruisers from Germany. He also purchased two steel-protected cruisers from Britain. At the same time, Japan was purchasing more warships from Britain and France, although on a smaller scale: two protected cruisers from France (both were 4150 tons) in 1890, two small cruisers in 1888 and one steel-protected cruiser from Britain and a state-of-the-art destroyer from France in 1892.

In reality, purchased ships were technologically more advanced and cost effective than Fuzhou-built ships. The *Beiyang* fleet's two steel-armored cruisers, the *Jingyuan* and the *Laiyuan* (completed in 1887 by the German shipbuilder, Vulcan) were 2900 tons (270.3 × 39.3 × 16 ft), equipped with a 5000 horsepower engine and cruised up to 15 knots. They were armed with two 8.2-inch and two 5.9-inch guns and protected by 9.5-inch thick above-water belt armor. Each one of these two cruisers cost 865,000 taels.[86] The *Nanyang*'s slightly more humble steel-hulled cruisers, the *Nanrui* and the *Nanchen* (completed in 1883 by the German shipbuilder, Howaldt) were even less costly. They were 2200 tons (275.5 × 37.5 × 18 ft) and equipped with 2400 horsepower engines that reached 15 knots. They were armed with two 8.2 inch and eight 4.6-inch guns and were roughly the same class as the Fuzhou-built *Pingyuan*, the best warship Fuzhou Navy Yard had ever built. But they cost only 270,000 taels each.[87]

In contrast, Fuzhou-built ships would not have been cost effective. First of all, the building costs of the Fuzhou vessels were high. The composite cruiser *Kaiji* cost 386,000 taels and the steel-hulled *Pingyuan* cost 524,000 taels.[88] They were more expensive than but technological inferior to the imported ships. Besides, the Navy Yard itself was a costly enterprise. Between 1866 and 1889, the Qing government and invested 12 million taels into the navy yard, which only managed to build 30 vessels, most of which were wooden-hulled gunboats or transports. Even, the *Pingyuan*, the best ship Fuzhou could build by the 1890s, could hardly compete with foreign-built warships.[89] Qing senior officials would have known that it would take more money and more time for the Fuzhou Navy Yard to reach the technical capacity to build vessels similar to the German-built *Jingyuan*. Yet, shipbuilding technology in the West was still advancing in the 1870s and the 80s. Therefore, senior officials chose purchasing as the way of building up the Qing naval forces. The imperial court in effect supported the policy by injecting money into the Maritime Defense Fund. The change in mood would have been perceived by the Fujian provincial government, which was given the responsibility of providing funds to the Fuzhou Navy Yard. As a consequence, the Fujian authorities would have preferred fulfilling other tax obligations to supporting the Fuzhou Navy Yard.

The decline of the navy yard was also reflected in the declining status of its director. Shen Baozhen's immediate successor, Ding Richang, former governor of Jiangsu and one of Li Hongzhang's most prominent protégés, was also concurrently governor of Fujian, a position that could have allowed him to mobilize the provincial financial resources.[90] Unfortunately, Ding left the position within a year without any major achievement. After Ding, none of the directors attained a similar level of political prestige even though some of them had close relations with Li Hongzhang. After the early 1880s, faced with deficits, the directors were forced to consider cutting down the labor force of the navy yard and keeping only the workmen essential for servicing steamships. Some of them proposed to expand the navy yard and applied for more funds.[91] Yet, none of these suggestions brought about any real actions.[92] In other words, in the eyes of Qing officialdom and the imperial court, the Fuzhou Navy Yard had lost it status as one of the most important government undertakings for fulfilling the self-strengthening policy.

The result of the Sino-French War had fully exposed the weakness of the Fuzhou Navy Yard's technology when it was challenged by much more advanced French ironclads and firepower. Impatient senior officials did not see the value of investing more money in the navy yard, although the navy yard's engineers were eager to upgrade the navy yard's technical capacity. The *Pingyuan* project shows how such an opinion could make a negative effect on the navy yard's plans. After the war, the Fuzhou Navy Yard proposed to the imperial court, with Zuo Zongtang's backing, to build three steel-hulled steel-armored cruisers based on a French design with a budget of 400,000 taels for each vessel.[93] The imperial court gave its permission.[94] However, Li Hongzhang, who was planning to purchase two steel-armored cruisers from Germany, strongly criticized the Fuzhou proposal by arguing that the proposed ships would be weaker than the ironclads he had purchased, hence spending money on such a project would not be wise. He requested the imperial court to reconsider the decision.[95] Unfortunately, Zuo had died just after he gave support to the new project. Without its principal political backing, the navy yard was unable to defend itself against Li's criticism. The imperial court Consequently, the imperial court only granted 460,000 taels to be provided by the Fujian authorities to build one of the three proposed ships.[96]

Therefore, the result of the directors' lack of political weight and ability to persuade senior officials at the imperial court was that funds were never given to the navy yard for upgrades. After 1885, only Zhang Zhidong, governor-general of Guangdong and Guangxi, who had been interested in building up a navy, provided financial support for Fuzhou's shipbuilding works. He ordered eight warships from the navy yard. Unfortunately, In October 1889, Zhang was appointed governor-general of Hunan and Hubei. His successor in Guangzhou suspended his predecessor's orders. Hence, the navy yard only received 354,545 taels for half of the building costs from Guangdong. Consequently the shipbuilding works on one ship dragged on for four years. This dire situation only changed after the Sino-Japanese War, when the imperial court and senior officials discussed reviving the Fuzhou Navy Yard and, once again, appointed the governor-general of Fujian and Zhejiang as director of the navy yard.

Conclusion

Between the 1870s and 1880s, the Fuzhou Navy Yard was trapped in the awkward position of having to prove its effectiveness before it could become effective. Like all new industrial enterprises, the Fuzhou Navy Yard had to overcome the learning curve to prove its worth, and, given the demand for a high quality end product, was given relatively little time to accomplish the required standard. It started by giving away its ships, but, even then, the provinces were reluctant to pay the cost of their maintenance and, therefore, reluctant to accept them. Ultimately, the arms race between China and Japan, and the realization that Japan was buying rather than building its best warships, persuaded senior Qing officials that they, too, had little choice but to substitute buying for building. The purchases competed with the Fuzhou Navy Yard for the same financial sources, the maritime customs. In such a situation, the Qing government's fiscal weakness was exposed. Its rigidness in budgeting and the system of tax quotas made the Fujian authorities find it difficult to fulfill all their fiscal responsibilities. Hence, when officials sensed the shift of naval building policy from building to buying, they chose to leave the Fuzhou Navy Yards aside but to support other financial needs. Furthermore, the Fuzhou Navy Yard was unable to elevate its political status. Hence, the consequence was that the Fujian authorities contributed less and less money to the navy yard to a point that the navy yard could barely expand and upgrade its facilities. The few warships that it built in the 1870s and the 1880s marked its technical progress. Yet the lack of funds finally halted its progress in the 1890s.

It might be said that other weaknesses included the fragmentation of authority in the Qing government's bureaucratic structure and a technological environment that, in the 1870s, was not yet tuned to the rapid technological change. Indeed, it might have been a shortcoming in shipbuilding that the creation of a modern navy was left to provincial efforts. In a more efficient setting, the navy yard should not have had to persuade individual provinces to accept its ships, but some overall naval defense with better appreciation for the development of modern ships over junks. Individual provincial officials, Zuo Zongtang, Li Hongzhang, and Shen Baozhen included, should not have had to continually memorialize that funds were not delivered.[97] The Maritime Defense Fund might have received the full 4 million annual budget, more than the actual around 2.1 million taels per year (42 million in total) between 1875 and 1894.[98] The lack of a centralized navy, with full realization that China needed its own navy yard, made it much harder for the Fuzhou Navy Yard to obtain funds, and therefore, the long-term investment for technological development.

However, it is not true, as historians have argued, that navy yard personnel preferred a civil service career to a technical career, and that preference hindered the Fuzhou Navy Yard's progress.[99] The Fuzhou Navy Yard was always staffed by engineers (foreign or Chinese) who could keep up with the progress of Western technology and, though not always, earn the director's support in launching new projects. Engineers could bear imperial official titles and develop a career in government projects and even as government officials. Hence, proving the effectiveness of the navy yard would be a strong case for promotion. Shen Baozhen's execution of the five-year contract as well as his handling of the Taiwan Crisis, for example, had earned him credentials in the self-strengthening politics. Such credentials were the stepping-stone for him to be promoted to the position of governor-general of Jiangxi, Jiangsu, and Anhui, which was the second most senior position in the late Qing local officialdom. The core problem was how the central government and senior officials viewed the impact of the navy yard. The relatively less successful career of Shen's successors in the Fuzhou Navy Yard might be the consequence of their inability to prove their own competence, and the high turnover of the candidates after Shen (three resigned and four moved to other mid-ranking positions between 1875 and 1890) shows that they either could not handle the politics or the imperial court did not consider them worthy of directing the navy yard.

The central government's attitude changed after the Sino-Japanese War. In 1896, after reviewing the Fuzhou Navy Yard, the imperial court appointed Yulu, the Fuzhou Manchu General, who was one of the closest allies of the Empress Dowager Cixi, as director of the Fuzhou Navy Yard and, at the same time, governor-general of Fujian and Zhejian. The government intended to issue shares to attract Chinese merchants to join the navy yard, but the attempt failed.[100] The imperial court ordered Yulu to revive the shipbuilding program, to open coal and iron mines in Fujian, and to make up the accumulated arrears for the past years. Yulu recruited a group of five French technicians on the basis of a five-year contract, and tried to deliver as much money as he could to the navy yard, though not the full amount. With the improved financial support, between 1896 and 1905, the navy yard completed seven steel-hulled ships. Later five ships were built under French supervision. Unfortunately, the available documents reveal little about the technical progress of that period. Two of the five ships were steel-armored torpedo gunboats that were equipped with 6500 horsepower triple expansion engines whose top speed reached 23 knots (870 tons, 260 × 26.6 × 11 feet). They were completed in 1903. Yet, in 1905, because of a dispute over a French order to build merchant ships, the Qing government dismissed the French technicians and suspended the shipbuilding program.

To make the Fuzhou Navy Yard a more efficient navy yard, the Qing government had to consider a wide-range of military reforms able to accommodate shipbuilding as part of the military development. Traditional navies had to be disbanded, war junks decommissioned, and seamen retrained. Clear strategic and technological thinking on naval development via both building and purchases needed to be drawn up, and then key technologies had to be transferred into the navy yards from leading countries of naval technology. The British Royal Naval Dockyards, one of the leaders in naval technology in the nineteenth century, went through a similar process. Technological development was set as a priority since it would be translated into military strength. They were free from competition and hence were usually criticized as less efficient and costly than private dockyards. However, backed by the British Admiralty, they were pioneers in key naval technologies such as the use of iron and progressive versions of the boiler. The Admiralty had a department directing naval navy yards. It also had the Royal School of Naval Architecture and Marine Engineering that trained a large number of engineers and technicians. Since naval dockyards did not have sufficient capacity to build all its ships, the Admiralty subcontracted shipbuilding works to private shipbuilders such as the prominent shipbuilding and marine engineering company Armstrong. Private shipbuilders could acquire new naval technologies by obtaining naval contracts and, on many occasions, recruiting engineers from the Royal Navy. Furthermore, competition drove private shipbuilders to develop new designs that might attract contracts from the Royal Navy or foreign governments.[101] That is to say, the demand for naval development drove both government and private dockyards into developing naval technology.

The Qing government navy yards could have played a similar role. Between the 1870s and 1880s, the Fuzhou Navy Yard was a pioneer in importing new technologies from the West with the backing from the government. It could have been one of the sources of technology for domestic shipbuilders had it been given enough funds to develop. In effect, after 1895, when the government needed to rebuild its naval force, despite the huge indemnities imposed by the Treaty of Shimonoseki and the Boxer Protocol, the Qing government revived the shipbuilding programs in the Fuzhou Navy Yard and the Jiangnan Arsenal. Unfortunately Fuzhou's program did not survive into the Republican era. The collapse of the political order and the 1911 Revolution finished off any semblance of central government direction and, hence, the central government's connection with the navy yard.

Notes

1. *Zuo Wenxiang gong quanji*, Memorials 18:2a-5b; *Haifang dang* II, pp. 6–8. *Yangwu yundong*, vol.5, pp. 445–446.
2. *Haifang dang* II, p. 7.

3. *Xianfeng Tongzhi liangchao shangyudang*, vol. 16. (Guilin: Guangxi shifan daxue, 1998), p. 216.
4. *Yangwu yundong*, vol.5, pp. 13–14.
5. *Zuo Wenxiang gong quanji*, Memorials, 19:28a.
6. *Yangwu yundong*, vol.5, p. 15; *Xianfeng Tongzhi liangchao shangyudang*, vol.16, p. 276; *Yangwu yundong*, vol.5, p. 17, *Haifangdang*, II, p. 22. It is worth noting that the imperial court did not appoint Shen Baozhen as imperial commissioner. The post of *Chuanzheng dachen* was not institutionalized until his successor Ding Richang was appointed in 1874.
7. Shen Baozhen was Lin Zexu's son-in-law. He became associated with Zeng Guofan, who was at that moment stationed in Nanchang, the capital of Jiangxi, when he was appointed prefect of Jiujiang, Jiangxi in 1856. With his own military achievement and Zeng Guofan's patronage, Shen climbed up the bureaucratic ladder quickly and was appointed as governor of Jiangxi in 1862. His governorship in Jiangxi was successful. Several Taiping leaders were arrested and executed in the province in 1864. However, during this period Zeng and Shen seem to have fallen apart because Shen refused to divert Jiangxi's tax revenue to support Zeng Guofan, who was then governor-general of Jiangsu, Jiangxi, and Anhui and the imperial commissioner in charge of the military affairs of Jiangxi, Jiangsu, Anhui, and Zhejiang. For more details, see David Pong, *Shen Pao-chen*, pp. 73–86. Zuo Zongtang admired Li Zexu whom he had only met him once. Zeng Guofan told his friend in a letter of late 1867 that Zuo and Shen had become "sworn confederates." Zeng Guofan, *Zeng Guofan weikan xingao* (Beijing: Zhonghua shuju, 1959), p. 391.
8. Zuo recommended five of his protégés to assist Shen Baozhen. They were Zhou Kaixi, Hu Guangyong, Wu Dating, Ye Wenlan and Huang Weixuan. Zhou Kaixi was on Zuo's staff. Hu Guangyong was a merchant-banker, who had been raising funds to support Zuo's campaign against the Taipings since the early 1860s. Wu Dating was formerly Taiwan *daotai*. *Yangwu yundong*, vol.5. pp. 33–34. For more details about the appointment of Shen Baozhen and the Fuzhou Navy Yard's directors, see David Pong, *Shen Pao-chen*, pp. 107–140.
9. *Haifang dang*, II, p. 6.
10. David Pong, "Keeping the Foochow Navy Yard Afloat," *Modern Asian Studies* 21:1 (1987), pp. 126–127.
11. When Li Hongzhang established it in 1865, the initial cost of purchasing the machine shop from an American firm, Thomas Hunt & Co., was paid by two maritime customs clerks, who were convicted of corruption. Its running costs, which were around 15,000 taels per month, were supported by Li's *Huai* Army budget.
12. *Zhongguo jindai bingqi gongye dang'an shilia*, vol. 2, p. 575.

13. *Zhongguo jindai bingqi gongye dang'an shiliao*, vol. 1, pp. 575–576; *Haifang dang*, III, pp. 55–56.
14. After Li became governor-general of Zhili in 1870, his protégés Ding Richang, Shen Bingcheng, and Feng Junguang, as successive Shanghai *daotai* managed both the Shanghai maritime customs' income and the arsenal. After Feng Jungang resigned from Shanghai in 1874, the relationship between the Arsenal general managers and Li Hongzhang were less clear. However, after 1884, Li Hongzhang still exercised enough influence over the arsenal to recommend promotion for its personnel. Stanley Spector, *Li Hung-chang and the Huai Army* (Seattle: University of Washington Press, 1964), pp. 131–133, 157, 164.
15. The other three directors were governor-general of Huguang, governor-general of Jiangsu, Jiangxi and Anhui, and governor of Jiangsu who could never reach Li's prominence before 1895. Li Hongzhang was governor of Jiangsu 1862–1867, acting governor-general of Jiangxi, Jiangxu, and Anhui, 1865–1866, and governor-general of Hunan and Hubei, 1867–1870.
16. Wei Yungong (compl.), *Jiangnan zhizaoju ji*, 4:2–4.
17. Although there is no evidence, Zuo Zongtang may have consulted the officials of the coastal provinces in private before making such a statement.
18. The term is first used in Zuo Zongtang's memorial of January 1885 to describe the Fuzhou steamboats built in the 1860s and 1870s. *Yangwu yundong*, vol. 5, p. 307.
19. Li Wenzhi and Jiang Taixin, *Qingdai caoyun* (Beijing: Zhonghua shuju, 1995), p. 439.
20. Ni Yuping, *Qingdai caoliang haiyun yu shehui bianqian* (Shanghai: Shanghai shudian, 2005), pp. 152–157.
21. In 1867, Rong Hong submitted a proposal to Zeng Guofan for forming such a company. Rong Hong told Ding Richang that the steamers of the proposed company could be used in shipping tributary rice. Zhang Guohui, *Yangwu yundong yu Zhongguo jindai qiye* (Beijing: Zhongguo shehui keshue, 1979), pp. 129–134; Peng Deqing (ed.), *Zhongguo hanghai shi (Jindai hanghai shi)* (Beijing: Renmin jiaotong chubanshe, 1989), pp. 112–114.
22. *Li Wenzhonggong quanji*, Letters to Friends and Colleagues, 11:31b.
23. *Yangwu yundong*, vol. 5, p. 87, 90. Some other 150-horsepower Fuzhou boats could carry even less cargo. Such as the fourth boat could only load 560 tons, and the fifth only 400 tons. An 80-horsepower Fuzhou boat cost 1,178 taels monthly. *Yangwu yundong*, vol. 5, p. 96, 102.
24. I cannot find a figure for the coal consumption of the Fuzhou vessels. But according to the Fuzhou Navy Yard's financial report of August/September 1874–January 1877, the 10 Fuzhou steamers spent 63,605

taels on coal. Hence a steamer could have consumed 200 taels a month. *Yangxu yundong* vol. 5, pp. 232.
25. In 1886, the Qing government paid 0.4 taels for shipping a *shi* of rice, hence a 3,000-*shi* junk could earn around 1,200 taels. The Board of Revenue, "Fuchen weichi zhaoshangju shiyi shu," in Ge Shijun (ed), *Huangchao jingshiwen xubian*, 113:8b.
26. *Haifang dang*, II, pp. 166–168; *Yangwu yundong*, vol. 5, pp. 85–86. This opium tax was raised to pay the costs of purchasing guns and steamships in 1862. After the third month of 1867, the Board of Revenue instructed the Fujian government to deposit the tax and wait for further instruction.
27. It was around 5,200 taels per lunar month for 15 lunar months. David Pong, "Keeping the Foochow Navy Yard Afloat," *Modern Asian Studies* 21:1 (1987), p. 128.
28. Zhejiang took the second Fuzhou steamer. Guangdong did not take over any but had been building its own naval force by purchasing steamers from abroad since 1866. Ruilin, governor-general of Guangdong and Guangxi purchased six gunboats from France and Britain in 1866. *Chouban yiwu shimo*, Tongzhi reign, 55:5a-6b.
29. *Yangwu yundong*, vol. 5, p. 103.
30. Coastal provinces' responses were mixed. In the discussion, Zhejiang had one Fuzhou gunboat in service but intended to replace the Fuzhou one with a newer Fuzhou vessel. Shandong and Fengtian expressed their difficulty of raising funds for supporting steamers and asked for permission to divert money from the maritime customs. Li Hongzhang considered larger Fuzhou vessel's deep draught and hoped to take over a smaller 80-horspoer vessel. And Ruilin, governor-general of Guangdong and Guangxi quickly agreed to take over one 150-horsepower Fuzhou steamer because it considered it effective in rough sea conditions. *Haifang dang*, II, pp. 317–318, 329, 337–339, 341, 344–345.
31. Li Hongzhang took over the sixth Fuzhou steamer, Shandong decided to take over the eighth steamer, Zhejiang was to take over the fourth Fuzhou steamer to replace the second steamer. Guangdong took the fifth, and the second steamer was to be dispatched to Fengtian. *Haifang dang*, II, pp. 306–307, 311–312, 333–433, 396–397, 399, 401.
32. David Pong, "Keeping Foochow Navy Yard Afloat," p. 132.
33. By April 1872, Li Hongzhang had used the *Wannian Qing* and the *Anlan* to ship rice to Tianjian from Shanghai, but, it seems to have been a one-off mission. *Li Wenzhonggong quanji*, Letters to Colleagues, 12:2a, 9a.
34. *Yangwu yundong*, vol. 5, pp. 119–124.
35. *Haifang dang*, II, p. 422, 442.

36. *Haifang dang*, II, p. 502.
37. In January 1873, the company took over the thirteenth Fuzhou vessel, which was built according to the company's technical requirement. Yet, the company returned it to the navy yard in September and took over the fourteenth and fifteenth *Haifang dang* II, p. 853, 864; Li Zhigang, "Lunchuan zhaoshangju jingying guanli wenti," *Zongyang yanjiyyuan jindaishi yanjiusuo jikan* 19 (1990), p. 78; Zhongguo hanghai xuehui (ed), *Zhongguo hanghai shi (Jindai hanghai shi)* (Beijing: Renmin jiaotong chubanshe, 1989), p. 134.
38. *Haifang dang*, II, p. 410.
39. *Yangwu yundong*, vol. 5, p. 154.
40. *Haifang dang*, II, p. 530.
41. David Pong, "Keeping the Foochow Navy Yard Afloat," p. 134.
42. *Yangwu yundong*, vol. 5, pp. 154–162.
43. Li Hongzhang's suggestion was that provincial governments should take over ships from the two navy yards, decommission traditional war junks and disband water forces, and move the funds to support steamships. The *Zongli yamen* agreed with him and called for discussion on the issue of decommissioning traditional war junks. Yet, no discussion followed. *Yangwu yundong*, vol. 5 pp. 126–128.
44. *Haifang dang* II, p. 496. Li Zongxi, governor-general of Jiangsu, Jiangxi, and Anhui considered that China only needed between 10 and 20 steamboats of 600 tons to 1000 tons to ship tributary rice.
45. In the nineteenth century, mining coal with machinery meant employing the steam engine to pump water out of the mine shaft and to hoist the workers and coal up to the ground. Large scale mines may have used steam locomotives to transport coal. All the digging work was carried out by using hand tools. Mechanization progressed very slowly in the mining industry.
46. *Yangwu yundong*, vol. 5, p. 135, 137.
47. For details of the Qing fiscal system, see Yeh-chien Wang, *Land Taxation in Imperial China, 1750–1911* (Cambridge, Mass.: Harvard University Press, 1973), pp. 8–19; Marianne Bastid, "The Structure of the Financial Institutions of the State in the Late Qing," in Stuart R. Schram (ed.), *The Scope of State Power in China* (London: School of Oriental and African Studies, 1985), pp. 51–79.
48. The Kangxi emperor famously decreed that "*shengshi zisheng rending yongbu jiafu* (no taxes for new born babies)" in 1712. For Qing emperors, it was also a popular political move to cancel certain tax quotas. *Daqing huidian shili* (Supplementary precedents and regulations of the Collected Statues of the Great Qing), 265:10–12, quoted in He Lie, *Qing Xiantong shiqi de caizheng* (Taipei: Guoli bianyi guan, 1981), p. 45.

49. The Kangxi emperor campaigned against the Three Feudatories (1673–1681) and Taiwan (1683). The Qianlong reign (1736–1796) launched a series of expeditions to central Asia, Western Sichuan, Burma, and Vietnam.
50. Zhou Zhichu, *Wanqing caizheng jingji yanjiu* (Jinan: Qilu chubanshe, 2002), pp. 64–67.
51. In 1853, Lei Yicheng, vice president of the Board of Punishment was sent to Yangzhou to assist military affairs. He raised funds locally by initiating duty on rice in the nearby market towns. In the next year, the court agreed to institute the duty and extend it to cover not only rice but other merchandise. Other provinces soon followed suit. A detailed discussion of the impact of the Taiping Rebellion on the fiscal system, see, Zhou Zhichu, *Wanqing caizheng jingji yanjiu*, pp. 63–99.
52. Zhou Yumin, *Wanqing caizheng yu shehui bianqian* (Shanghai: Shanghai renmin, 2000), p. 238.
53. A more detailed discussion on the devolution of the Qing fiscal system can be found in Zhou Yumin, *Wanqing caizheng yu shehui bianqian*, pp. 222–233; Zhou Zhichu, pp. 68–107; Chen Feng "Qingdai zhongyang caizheng yu defang caizheng de tiaozheng," *Lishi yanjiu* No. 5 (1997), pp. 100–114.
54. Chen Feng, "Qingdai caizheng zhichu zhengce yu zhichu jiegou de biandong," *Jianghan luntan* No. 5 (2000), pp. 72–73.
55. *Yangwu yundong*, vol.3, p. 541. The *Guangxu kuaiji lu* shows similar figures, see Zhou Zhichu, pp. 153–154.
56. Kwang-ching Liu and Richard J. Smith, "The Military Challenge: the North-west and the Coast," in John K. Fairbank and Denis Twitchett (eds), *The Cambridge History of China*, vol. 11, p. 239. The Chinese historian Zhou Yumin's calculation is slightly higher at 58.5 million taels. See Zhou Yumin, *Wanqing caizheng yu shehui bianqian*, p. 266.
57. Zhou Yumin, *Wanqing caizheng yu shehui bianqian*, p. 268.
58. *Yangwu yundong*, vol.5, p.21.
59. From the Fujian maritime customs, the Fuzhou Navy Yard had been allotted 400,000 for capital construction, since the 12th month of 1866 to the twelfth lunar month of 1871, the aggregation of the 500,000 of monthly remittance had reached 3,150,000 tales. *Haifang dang*, II, p. 330, *Yangwu yundong*, vol.5, p.107.
60. *Yangwu yundong*, vol. 5, pp. 106–107.
61. Japan purchased two previously-owned ironclads in 1867 and 1870. They were both built in 1864 for the Confederate States during the American Civil War. The Fujian authorities received a junior military officer's intelligence report that Japan had such vessels. This discovery was a surprise, although they knew that Japan had been purchasing Western

armament since the 1860s. They considered that China might be able to defend against the Japanese expedition force to Taiwan, but the ultimate result would be unpredictable if Japan launched a large-scale assault against China. Hence, they thought to avoid a full-frontal confrontation and, in October 1874, China agreed to compensate the Ryukyuan victims and the costs of the Japanese expedition. *Chouban yiwu shimo*, Tongzhi reign, 93:47a, 94:4b, David Pong, *Shen Pao-chen*, p. 292. John Rawlinson, *China's Struggle for Naval Development*, p. 61. For more details on the Taiwan Incident, see Leonard Gordon, "Japan's Abortive Colonial Venture in Taiwan, 1874," *The Journal of Modern History* 32:2 (1965), pp. 171–85; Immanuel C.Y. Hsu, "Late Ch'ing Foreign Relations, 1866–1905," in Denis Twitchett and John K. Fairbank (eds.), *The Cambridge History of China* vol.11, (Cambridge: Cambridge University Press, 1980), pp. 85–88.

62. In the discussion, Li Hongzhang argued that China should suspend the costly north-western expedition and move the funds to support maritime defense. This idea brought the opposition from those senior officials who believed that Russia also posed a serious threat to China and frontier defense should be equally important to maritime defense. For details about the debate on maritime defense of 1875, see Immanuel C.Y. Hsü, "The Great Policy Debate in China, 1874: Maritime Defense V.S. Frontier Defense," *Harvard Journal of Asiatic Studies* 25 (1964–1965), pp. 212–228.

63. The elongated explosive shell was invented by Henri-Joseph Paixhans, a French artillery officer, in 1822. For the development of the explosive shell and a detailed study on the birth of the ironclad, see James Phinney Baxter, *The Introduction of the Ironclad Warship* (Cambridge, Mass.: Harvard University Press, 1933), pp. 17–32; Bernard Brodie, *Sea Power in the Machine Age* (Princeton: Princeton University Press, 1943), pp. 171–265.

64. The French navy built heavily armored floating batteries that fought against Turkish wooden battleships with success. After the war, the French started to build ironclads. They were essentially wooden steam battleship plated with 4 1/2 inch armor and extensive iron structural reinforcement and an iron deck. After the Crimean War, the French navy built a fleet of ironclads and the most famous one was the *Glorie*, built in 1859. Britain answered the French challenge with iron-hulled and 4.5-inch armored battleships, the *Warrior* and the *Black Prince*, in 1861. This was the beginning of the age of the ironclad.

65. In 1871 a French ironclad visited Shanghai. Feng Junguang the Shanghai *daotai*, boarded it and gave Li Hongzhang a rough description. The ship was a two-decker ship and the hull was roughly 5.5 inch plank. It was

armored by roughly 3.2 inches iron plates. The gun decks and the engine were protected by more iron plates. *Yangwu yundong*, vol.5, p. 121.
66. *Yangwu yundong*, vol.1, p. 48, 57, 63, 67, 72–73, 78–79, 84, 86, 89, 97, 108.
67. *Yangwu yundong*, vol.1, p. 49; 162–165. The transit dues revenue of the provinces of Jiangsu, Zhejiang, Hubei, and Jiangxi, and the maritime customs income of Jiangsu, Zhejiang, Fujian, and Guangdong shared the duty of delivering funds to the Maritime Defense Fund. With the establishment of this Fund, senior officials started to purchase warships from Germany and Britain. See John Rawlinson, *China's Struggle for Naval Development*, pp. 67–79; David Pong, "China's Defense Modernization and the Revenue of the Maritime Customs Service, 1875–1879," in Yen-p'ing Hao and Hsiu-mei Wei (eds), *Tradition and Metamophosis in Modern Chinese History: Essays in Honor of Professor Kwang-ching Liu's Seventy-Fifth Birthday* (Taipei: Institute of Modern History, Academia Sinica, 1998), pp. 980–981.
68. This was from Shen Baozhen's memorial of 20th July 1875. *Yangwu yundong*, vol. 5, p. 167.
69. The Fujian authorities had memorialized concerning the difficulties of the Fujian maritime customs and, in December 1874, the court instructed the Board of Revenue to draw up a plan. *Yangwu yundong*, vol.5, pp. 163–164.
70. *Haifang dang*, II, pp. 550–551.
71. *Yangwu yundong*, vol. 5, pp. 166–167.
72. *Haifang dang*, II, p. 644.
73. *Yangwu yundong*, vol. 5, p. 242.
74. For more details of Li Hongzhang's preference for purchased warships, see, John Rawlinson, *China's Struggle for Naval Development*, pp. 63–95.
75. *Li Wenzhonggong quanji*, Letters to Friends and Colleagues, 12:3b.
76. *Yangwu yundong*, vol.1, pp. 40–54.
77. In 1879, Li urged the Fuzhou Navy Yard to carry out the composite cruiser project without helping to obtain any money. In January 1881, he suggested that building ironclads was advisable and urged the navy yard to draw up a budget—a suggestion that may have had to do with the returned naval engineers. *Li Wenzhong gong quanji*, Memorials, 39:30a–31b; *Yangwu yundong*, vol.5, p. 188.
78. *Yangwu yundong*, vol.5, p. 216.
79. These two vessels could be the "ironclads" that alarmed Qing senior officials during the Taiwan Crisis in 1874. David Evans and Mark Peattie, *Kaigun: Strategy, Tactics, and Technology in the Imperial Japanese Navy, 1887–1941* (Annapolis, Naval Institute Press, 1997), p. 14.

80. These gunboats were designed by the British naval architect, George Rendel. They were around 250–320 tons, equipped with 180 horsepower engine, around 120 × 30 × 9 feet dimension, but mounted a single 10-inch gun. Shen Baozhen bought two iron gunboats during the Taiwan Crisis. Li Hongzhang purchased four similar boats in 1875. Later Shen ordered four when he was in Nanjing in 1878. Those boats were taken over by Li in 1879 when they arrived. Three similar gunboats were ordered in 1880. For more details of these purchases, see John Rawlinson, *China's Struggle for Naval Development*, pp. 69–71.
81. The protected cruiser means that only the machinery is protected by the armour under the deck. The armored ship had a belt-armor along its side.
82. After Shen Baozhen's death in 1879, in April 1880, the Fuzhou Navy Yard and the Fujian authorities submitted a joint memorial, requesting that the imperial court instruct the newly appointed imperial commissioner of the southern ports to allocate funds for building the cruiser. With the advice from the Board of Revenue, the imperial court agreed that the navy yard should receive 63,000 taels, which was only a fraction of the 400,000 the Fuzhou Navy Yard needed. *Haifangdang*, II p. 856.
83. *Yangwu yundong*, vol. 5. pp. 265–266.
84. In February 1881, a prominent official who had also distinguished himself by fighting the Taipings much earlier, memorialized for instructing the Fuzhou engineers to build ten small gunboats that could be employed on the Yangzi River. Unfortunately no evidence suggests that the project was carried out. *Haifangdang*, II, p. 872.
85. In November 1882, Zuo Zongtang memorialized to build four more cruisers. They also pushed the navy yard to build the ten gunboats that Peng Yulin proposed earlier on. Li Hongzhang also ordered two cruisers for his *Beiyang* fleet. *Zuo Wenxiang gon quanji*, Memorial 59:52ab; *Haifangdang*, II., p. 938.
86. The *Beiyang* fleet's capital ships, the *Dingyuan* and the *Zhenyuan* (completed in 1884 by the German shipbuilding, Vulcan) were 7,670 tons (dimension: 308 × 60 × 20 ft). They were equipped with 7,200-horsepower engines that could reach 15.4 knots. The midship armor was 14-inch thick and they were armed with four 12-inch and two 5.9-inch guns. They cost around one million taels each. The *Beiyang*'s two steel-protected cruisers, the *Zhiyuan* and the *Jingyuan* (completed in 1887 by Armstrong) were 2300 tons (268 × 38 × 15 ft) and equipped with 6850 horsepower engines that could reach 18.5 knots. Each one of them cost 845,000 taels.
87. Some senior officials claimed that Fuzhou composite cruisers were better than the *Nanrui* and the *Nansheng*. See Liu Kunyi, *Liu Zhongchenggong yiji*, Memorials, 24:43.

88. Zhang Guohui, *Yangwu yundong yu jindai qiye*, p. 49.
89. John Rawlinson, *China's Struggle for Naval Development*, p. 145.
90. The historian John Rawlinson suggests that Ding resigned to avoid dealing with the financial problem. John Rawlinson, *China's Struggle for Naval Development*, p. 100.
91. *Yangwu yundong*, vol. 5, pp. 293–294; *Chuanzheng zhouyi*, 24:1a–5b.
92. A more controversial spending was the Empress Dowager Cixi's project to build the summer palace. The estimated cost of building was 10 million taels in total between 1885 and 1895, which was on average one million per year. It is estimated that the summer palace project had obtained 7,500,000 to 10,000,000 taels loan from the Admiralty. Hence, it may be argued that such spending could have affected the Admiralty's work in organizing China's navies. For the estimated building cost of the summer palace, see Jiang Ming, *Longqi piaoyang di jiandui* (Shanghai: Shanghai Jiaotong daxue, 1991), p. 208. For the Maritime Defense Fund, see Zhou Yumin, *Wanqing caizheng yu shehui bianqian*, pp. 264–276. Another important public work was Yellow River conservancy, which cost millions annually. See Chen Hua, "Qingdai de hegong yu caizheng," *Qingshi yanjiu* No.3 August 2005, p. 35.
93. *Yangwu yundong*, vol. 5, pp. 311–313.
94. Zhang Xia et al. (ed.), *Qingmo haijun shiliao* vol.1 (Beijing: Haiyang chubanshe, 1982), p. 115.
95. *Li Wenzhonggon quanji*, Letters to the *Zongli yamen*, 17:40ab; Jiang Ming, *Longqi piaoyang de jiandui – Zhongguo jindai haijun xingshuai shi* (Beijing: Sanlian, 2002, revised edition), pp. 245–246.
96. *Yangwu yundong*, vol. 5, pp. 412–416.
97. *Yangwu yundong*, vol. 5, pp. 378–379, 424–425. Zuo Zongtang's way of ensuring funds was to obtain loans from foreign banks and ask coastal provinces to pay the loan from the revenues of the maritime customs. This idea could have had something to do with his close associate, Hu Guangyong, a banker who had good connections among foreign and Chinese merchants in Shanghai and had been the major fundraiser of his forces since the anti-Taiping campaign in the late 1860s. By obtaining foreign loans with imperial permission, Zuo could ignore the problem of arrears while provincial governments would still have to share the burden of both the campaign and the interest on the loans. For more details about the loans for Zuo Zongtang's campaign, see Xu Yi et al., *Qingdai waizhai shilun* (Beijing: Zhongguo caizheng, 1996), pp. 217–236.
98. The Chinese historian Jiang Ming estimates that the *Beiyang* had received 23,022,422 taels and spent 21,357,977 taels. Jiang Ming, *Longqi piaoyang di jiandui* (Shanghai: Shanghai Jiaotong daxue, 1991), pp. 136–137.

99. Mary Wright, *The Last Stand of Chinese Conservatism: The T'ung-Chich Restoration, 1862–1874* (Stanford: Stanford Univeristy, Press, 1957), p. 248; Knight Biggerstaff, *The Earliest Modern Government Schools in China*, pp. 71–84, John Rawlinson, *China's Struggle for Naval Development*, p. 104.
100. *Chuanzheng zouyi*, 46:16a.
101. For the relationship between the British Royal Navy and private shipbuilders, see Sidney Pollard and Paul Robertson, *The British shipbuilding Industry*, pp. 205–222.

References

Baojun 寶鋆 et al. (comp.), *Chouban yiwu shimo* 籌辦夷務始末, Tongzhi reign 同治朝 (Beipeing: National Palace Museum, 1930, reprint of 1880 edition).

Bastid, Marianne, "The Structure of the Financial Institutions of the State in the Late Qing," in Stuart R. Schram (ed.), *The Scope of State Power in China* (London: School of Oriental and African Studies, 1985), pp. 51–79.

Baxter, James Phinney, *The Introduction of the Ironclad Warship* (Cambridge, Mass.: Harvard University Press, 1933).

Brodie, Bernard, *Sea Power in the Machine Age* (Princeton: Princeton University Press, 1943).

Chen Feng 陈锋, "Qingdai caizheng zhichu zhengce yu zhichu jiegou de biandong 清代财政支出政策与支出结构的变动" *Jianghan luntan* 江汉论坛 No.5 (2000), pp. 60–70.

Chen Feng 陈锋, "Qingdai zhongyang caizheng yu difang caizheng de tiaozheng 清代中央财政与地方财政的调整," *Lishi yanjiu* 历史研究 No.5 (1997), pp. 100–114.

Chen Hua 陳樺, "Qingdai de hegong yu caizheng 清代的河工与财政," *Qingshi yanjiu* 清史研究 No.3 August 2005, p. 33–42.

Evans, David, and Mark Peattie, *Kaigun: Strategy, Tactics, and Technology in the Imperial Japanese Navy, 1887–1941* (Annapolis, Naval Institute Press, 1997).

Ge Shijun 葛士濬 (ed), *Huangchao jingshiwen xubian* 皇朝經世文續編 (Taipei: Wenhai chubanshe, 1972, reprint of the 1888 edition).

Gordon, Leonard, "Japan's Abortive Colonial Venture in Taiwan, 1874," *The Journal of Modern History* 32:2 (1965), pp. 171–185.

He Lie 何烈, *Qing Xian-Tong shiqi de caizheng* 清咸同時期的財政, (Taipei: Guoli bianyi guan, 1981).

Hsu, Immanuel C. Y., "Late Ch'ing Foreign Relations, 1866–1905," in Denis Twitchett and John K. Fairbank (eds.), *The Cambridge History of China* vol.11, (Cambridge: Cambridge University Press, 1980), pp. 85–88.

Hsü, Immanuel C. Y., "The Great Policy Debate in China, 1874: Maritime Defense V.S. Frontier Defense," *Harvard Journal of Asiatic Studies* 25 (1964–1965), pp. 212–228.
Institute of Modern History, Academia Sinica (compl.), *Haifang dang* 海防檔, (Taipei: Institute of Modern History, Academia Sinica, 1957). I *Goumai chuanpao* 購買船砲. II *Fuzhou chuanchang* 福州船廠. III *Jiqiju* 機器局.
Jiang Ming 姜鸣, *Longqi piaoyang de jiandui* 龙旗飘扬的舰队 (Shanghai: Shanghai Jiaotong daxue, 1991).
Li Hongzhang 李鴻章, *Li Wenzhong gong quanji* 李文忠公全集 (Jinlin: 1908 edition).
Li Wenzhi 李文治 and Jiang Taixin 江太新, *Qingdai caoyun* 清代漕运 (Beijing: Zhonghua shuju, 1995).
Liu Kunyi 劉坤一, *Liu Zhongchenggong yiji* 劉忠誠公遺集 (Beijing: Zhonghua shuju, 1959).
Liu, Kwang-ching and Richard J. Smith, "The Military Challenge: the North-West and the Coast", in John K. Fairbank, *The Cambridge History of China*, Vol.11 late Ch'ing (Cambridge: Cambridge University Press, 1980), pp. 202–273.
Mary C. Wright, *The Last Stand of Chinese Conservatism: The T'ung-Chih Restoration, 1862–1874* (Stanford: Stanford University Press, 1957).
Ni Yuping 倪玉平, *Qingdai caoliang haiyun yu shehui bianqian* 清代漕粮海运与社会变迁 (Shanghai: Shanghai shudian, 2005).
Peng Deqing 彭德清 (ed.), *Zhongguo hanghai shi (Jindai hanghai shi)* 中国航海史 (近代航海史) (Beijing: Renmin jiaotong chubanshe, 1989).
Pong, David, "China's Defense Modernization and the Revenue of the Maritime Customs Service, 1875–1879," in Yen-p'ing Hao and Hsiu-mei Wei (eds.), *Tradition and Metamorphosis in Modern Chinese History: Essays in Honor of Professor Kwang-Ching Liu's Seventy-fifth Birthday* (Taipei: Institute of Modern History, Academia Sinica, 1998), pp. 979–1006.
Pong, David, "Keeping the Foochow Navy Yard Afloat: Government Finance and China's Early Modern Defence Industry, 1866–75, *Modern Asian Studies*, 21:1 (1987), pp.121–152.
Pong, David, *Shen Pao-chen and China's Modernization in the Nineteenth Century* (Cambridge: Cambridge University Press, 1994).
Rawlinson, John L., *China's Struggle for Naval Development, 1839–1895*, (Cambridge, Mass.: Harvard University Press, 1967).
Spector, Stanley, *Li Hung-chang and the Huai Army: A Study in Nineteenth-century Chinese Regionalism* (Seattle: University of Washington Press, 1964).
Wang, Yeh-chien, *Land Taxation in Imperial China, 1750–1911* (Cambridge, Mass.: Harvard University Press, 1973).
Wei Yungong 魏允恭 (comp.), *Jiangnan zhizao ju ji* 江南製造局記 (Taipei: Wenhai chubanshe, 1969, reprint of 1905 edition).

Zhang Guohui 张国辉, *Yangwu yundong yu Zhongguo jindai qiye* 洋务运动与近代企业 (Beijing: Zhongguo shehui kexue chubanshe, 1979).

Zhang Xia 张侠 *et al.* (eds.), *Qingmo haijun shiliao* 清末海军史料 vol.1 (Beijing: Haiyang chubanshe, 1982).

Zhongguo jindai bingqi gongye dang'an shiliao weiyuanhui 中國近代兵器工業檔案史料委員會 (comp.), *Zhongguo jindai bingqi gongye dang'an shiliao* 中國近代兵器工業檔案史料, (Beijing: Bingqi gongye chubanshe, 1993).

Zhongguo shixuehui 中國史學會 (ed.), *Yangwu yundong* 洋務運動, 8 volumes (Shanghai: Shanghai renmin chubanshe, 1961).

Zhou Yumin 周育民, *Wanqing caizheng yu shehui bianqian* 晚清财政与社会变迁 (Shanghai: Shanghai renmin, 2000).

Zhou Zhichu 周志初, *Wanqing caizheng jingji yanjiu* 晚清财政经济研究, (Jinan: Qilu shushe, 2002).

Zuo Zongtang 左宗棠, *Zuo Wenxiang gong quanji* 左文襄公全集 (Taipei: Wenhai chubanshe, 1979, reprint of a late nineteenth century edition).

CHAPTER 7

Conclusion

In June 1893, the newspaper *North China Herald* reported on the contemporary state of river transport near Shanghai

> There must be hundreds of steam launches, many of them entirely constructed by Chinese, running on the inland waters in China, and yet it is very rarely that we hear of an avoidable accident occurring to them.

The article praised the Chinese people's technical ability saying that it was as good as any workmen in the world. The newspaper also commented on the capability of the Chinese managers of government machine shops and arsenals who had acquired the knowledge of machinery, asserting that, like their counterparts in Britain and America, their background of classical learning did not hinder their grasping the knowledge of machinery.[1] At around the time of the First Opium War in 1839, there would barely have been a single lathe or a steam engine on the China coast, except for what might already have been installed on a Western steamship. By 1895, in the treaty ports of Shanghai, Guangzhou and Hankou, there were around sixteen Chinese-owned machine shops that hired dozens of workers and employed machine tools, servicing and building small steam vessels.[2] Not spectacular perhaps, but the proliferation of machine shops signified China's transition to modern mechanical engineering.

True enough, by nineteenth-century standards, the technology of most Chinese machine shops did not constitute an industrial revolution. The age of steam was characterized by the steam engine, which China had had to learn to build, not in privately owned machine shops, but in government-funded and government-owned arsenals and shipyards. The introduction of the steam engine, too, had to come with a host of new science and technologies. Physics and mathematics had to be taught to young engineers. Coal, iron, and steel had to be worked by standards set in the West. The end products had to be the steamship and firearms, which Qing government officials were eager to obtain after the Second Opium War. Yet, it is all too easy to forget, when one sets eyes only on the products of warships and guns, that China started from such a very different techno-scientific foundation that the machine-minded did not realize it was thermal energy that drove the steam engine. It is also too easy to forget that the economics had to be right for new technology to survive. The Fuzhou Navy Yard required almost unlimited financial support in the face of growing threat which had been imposed by warships purchased from the West.

This book has set out to ask how, taking into consideration the intellectual and politico-economic realities of the time, steamship technology might have been transferred to China. It did not seek to assess if the technological record amounted to a success or a failure. To have done so would have grossly simplified the process of technology transfer and China's experience between the 1840s and 1890s. This book attempts to lay out a realistic record of the process, taking into consideration the complexities of the situation, and has sought to understand the decisions which were made in institutional terms. Institutional changes came slowly; but then, that was inevitable, given the very preliminary state of knowledge that Chinese people possessed of Western science and technology which they would acquire. By lodging the explanation for the decisions which were made in the transfer of technology in institutional terms, this book argues against the assumption, long resident in the secondary literature on the Self-strengthening Movement, that it was conservatism which killed the chance for success.

Institutional changes came slowly in the second half of the nineteenth century not because of conservatism. Although some people in the Qing officialdom did not accept the social and ideological implications of Western technology, their opposition was easily put down by the imperial court in the interest of military strengthening. Senior officials were also eager to exploit technology in pursuit of power and wealth. Yet, it could be argued, the Qing government had restricted itself by limiting foreign involvement in teaching science to young Chinese and, as in the case of the Fuzhou Navy Yard, dismissing most of the foreign technicians as a political gesture of intellectual and technical independence. However, the suspicion of foreigners in itself should not simplify the complexities of China's diplomatic situation. It is also significant that the substitution of foreign engineers was meant to be a demonstration that technological transfer had been achieved and that as Chinese managers took over at Fuzhou, Chinese engineers were still being trained in France. There might have been some misjudgments in the timing of these events—and no one in the 1870s and 1880s could have predicted the Sino-Japanese War and even less of its outcome—but there was no conservative triumph in stemming the import of Western technology.

What really held China back was the lack of well-trained technicians in new industry of steamship building. To push technological progress beyond the limit of a few government undertakings and foreign ventures, first of all, the education system had to be overhauled. By 1895, private dockyards and government undertaking that employed steam technology had only created an estimated 19,000 to 20,000 mechanics.[3] Yet, there was a bottleneck at the high-skill end, and to indigenize technology, China needed more than the few dozens of engineers that were trained by the Fuzhou Navy Yard. Training the highly skilled could not have been done by setting up a couple more new schools that had a modern engineering curriculum. By the end of the nineteenth century, apart from the *Tongwen guan* in Beijing and the *Guangfangyan guan* in Shanghai, there were only two new-style schools, in Guangdong and Hubei. Guangzhou's *Tongwen guan* was established as a language school in 1863 and no scientific subject was introduced. The other one was Zhang Zhidong's School of Self-strengthening (*Ziqiang xuetang*), established in Wuchang, Hubei in 1893. Originally its curriculum contained foreign languages, science (physics and chemistry), mathematics, and commerce. It intended to recruit 120 students at any time (30 for each subject). Yet, later in the same year Zhang Zhidong abandoned the subjects of science and commerce because of the lack of textbooks in Chinese. Hence, students only studied English, French, German and Russian languages and used Chinese traditional mathematics writings as textbooks.[4] Neither of those two schools could be considered engineering schools. New knowledge and skills had to come from somewhere else.

The case of the civil engineer Zhan Tianyou shows how difficult it was to train an engineer in the late Qing. Zhan followed Rong Hong's education mission to the United States in 1872 when he was 12 *sui*.[5] He entered the primary and high school in West Haven, Connecticut where he studied geometry, algebra, philosophy, and Latin. In 1878 he entered Sheffield Scientific School of Yale College, where he studied calculus, physics, geology, metallurgy, the steam engine, and civil engineering. He obtained the degree of Bachelor of Philosophy in civil engineering in 1881. After the education mission was withdrawn in the same year, he was sent to the naval school of the Fuzhou Navy Yard to learn navigation. After the Sino-French War, Zhan went to Guangzhou to teach English in a local school. In 1887, he joined the Kaiping Mine and took part in the construction of the mine's railway line between Tianjin and Tanggu under an English engineer.[6] Between 1890 and 1894 he joined the Qing government's railway and bridge building works in Manchuria. In 1895, he was elected as a member of the Institute of Civil Engineers in Britain. After that, his professional credibility was valued by senior officials and he conducted projects of railway construction around China, working side-by-side with foreign engineers.[7] Zhan Tianyou's case was rare. The small number of Chinese engineers highlighted the weakness of the Chinese education system in introducing Western technology on a large-scale.

As industrial late-comers, Japan and China's experience in transferring Western shipbuilding technology shared some similarities. For example, Japan's foremost government navy yard in the nineteenth century, the Yokosuka Navy Yard, was established in 1858 by the Satsuma government with technical assistance from a group of French naval engineers and technicians. It had a school that taught French, mathematics (basic, descriptive, and analytical geometry, algebra, and calculus), engineering drawing, and scientific subjects for two years. By 1871, it had a budget equivalent to the one-third of the annual budget of the Fuzhou Navy Yard. It started by building small wooden-hulled steamers but only built its first iron-hulled vessel in 1889. Yet, in the late 1880s, the yard had made technical progress, owing to the recruitment of a prominent French naval architect. It completed 13 warships between 1876 and 1894.[8] This number is far smaller than the Fuzhou Navy Yard's 33 built before 1895. The Japanese Imperial Navy, like the *Beiyang* fleet, heavily relied on purchased warships. In the Sino-Japanese War, only two out of 12 warships of the Japanese fleet were domestically built. Furthermore, lack of domestically-made steel and also lack of marine engineering capability made technical development in Yokosuka dockyard slow.

Nevertheless, the similarities stop here. When the Fuzhou Navy Yard dismissed most of its foreign technicians in the 1870s and struggled for funds in the 1880s, the Yokosuka Navy Yard recruited a prominent French naval architect in 1886 on the basis of a four-year contract and committed more funds to shipbuilding. He introduced the 4000-ton steel-hulled cruiser with a top speed of 18 knots to Yokosuka and the same design was also built in France. He directed the building of more cruisers, gunboats, and torpedo boats in Yokosuka, France, and Britain for the Japanese government.

More importantly, the Meiji government was interested in just establishing a few shipyards. Since the 1870s, new educational institutions outside the Yokosuka Navy Yard were opened. In 1873, the Imperial College of Engineering (*Kōbu Daigakkō*) in Tokyo was founded with the assistance of British engineers and teachers. Notably, those who had studied under William Rankine and William Thomson at the University of Glasgow were in charge of drawing up the curriculum. The college taught metallurgy, civil engineering, mechanical engineering, telegraphy, architecture, and chemistry.[9] In 1877, Tokyo University was established and incorporated the Imperial College of Engineering in 1886. In 1881, the Tokyo Vocational School was founded (becoming Tokyo Institute of Technology in 1929). These engineering schools trained engineers and technicians. Some of them gained scholarship to study abroad. For example, between 1878 and 1898, around 20 students studied at Glasgow with William Thomson, who became Lord Kelvin in 1892, in Glasgow, studying the most advanced physics of the time.[10] Although the number of those students was small, they went back to Japan to teach in universities and train more engineers who made contribution to the development of the shipbuilding industry Japan.

More institutional reforms were launched in the 1880s in Japan. Modern banks were set up in the form of a joint-stock company with limited liability. The Commercial Code was enacted in 1888. These changes helped to form large companies such as the Mitsubishi Gōshi Kaisha (becoming the prominent Mitsubishi Heavy Industry in 1937), which started from a shipping firm in 1873 and took over a government navy yard in Nagasaki in 1884. It adopted limited liability in 1886 and was incorporated in 1917. Japan's development of its shipbuilding industry in the twentieth century was not advanced by government navy yards but by private navy yards such as the Kawasaki Dockyard Co., which was incorporated in 1896, and the Mitsubishi's Nagasaki Navy Yard. These navy yards drew much from the expertise of foreign engineers and Japanese engineers who studied in state-established engineering schools or abroad. The technology progress of Japan in the early twentieth century highlights the need for institutional reform in China.

Therefore, for technology to be transferred, indigenized, and ultimately to achieve industrialization, China needed to produce many more engineers like Zhan Tianyou. Late Qing discussion of adding a few scientific topics to the imperial examinations would change little, because purpose of the imperial examinations itself was to select government officials, and the whole education system that followed that purpose downplayed fundamentals in science education. China needed engineering schools to be widely set up as well as changes made in the education system. Such educational reform was introduced when, in 1901, the Qing government instituted a series of reforms in which education was one of the essential parts. In 1905, the Qing government established the Ministry of Education (*Xuebu*) and, in the following year, decided to abolish the imperial examinations.

To be sure, the Qing government was short of funds even though funds were available in the country as a whole. The more serious problem was the lack of institutions for pooling capital, which made it difficult by the 1890s to even support the Fuzhou Navy Yard. By default, large-scale technological transfer in which the state saw a military interest had to be conducted directly by the Qing government. The post-Taiping Qing government, however, was more a conglomerate of provincial governments than a centrally directed bureaucracy, even though the imperial court retained, to the end, the right of appointment of senior officials. The dismantled railway from Shanghai, therefore, could not be rebuilt because the provincial officials in charge did not have he funds, and the Fuzhou Navy Yard had to persuade provincial authorities individually to take the navy yard's steamers, each of whom calculated in its own interest how costs might be kept to a minimum. The question of funding, therefore, leads directly to the question of whether the Qing government in the 1870s was structured in a way which allowed it to manage projects on a scale required for mechanized shipbuilding.

One of the reforms the government might have had to carry out, if government undertakings were to make a technological contribution, was allowing its engineering enterprises a freer hand in dealing with the market. Indeed, it can be argued that this was the direction which the late nineteenth-century Qing government had been taking. In the relatively successful case of the dockyard of the Jiangnan Arsenal, for example, in 1905, the governor-general of Jiangsu, Jiangxi, and Anhui decided to separate the dockyard from the arsenal, and recruited a team of British engineers to manage the dockyard. The dockyard was allowed to service and build ships for profit. It took orders from the China Merchants' Steam Navigation Company, foreign firms, the Chinese Maritime Customs, and the Chinese government. In the early Republican era, it became one of China's most important shipbuilders, although it did not build vessels larger than 3000 tons.[11] After the government gave the shipyard a free hand in its business ventures, it fared much better than the Fuzhou Navy Yard.

However, if profit is brought into the formula, as it must be, there is no reason why engineering enterprises should be left totally to government agencies. Examples of successful private engineering firms are plenty: Japan's Mitsubishi Goshi Kaisha went into steel making, ordnance, shipbuilding, engineering, and electricity, all outside the realm of government projects. However, to promote engineering firms of this scale, China needed institutions to pool capital. In late Qing China, foreign-owned dockyards in the form of joint-stock companies in Hong Kong and Shanghai had been expanding and making technical progress since the 1880s. Leading dockyards such as Boyd & Co. and Farnham & Co., both incorporated in the 1890s, had been taking orders from foreign merchants and the Chinese government, although they largely built smaller vessels for inland-river or coastal navigation.[12] The relatively flourishing foreign-owned industries in China highlighted Chinese entrepreneurs' slow progress, and the situation did not change until the Sino-Japanese War of 1894–95.

The first impact of the Sino-Japanese war on China's industrial enterprises was brought about by the Treaty of Shimonoseki, whose term removed the Chinese government's restriction on foreigners' right to build factories in treaty ports. The second impact came from the overhaul of the government which followed in the early years of the twentieth century. Central authority faded with the defeat in 1895 and subsequent events, especially the Boxer Uprising. The government which returned to Beijing from exile in 1901 was saddled with a substantial indemnity to pay its foreign victors, suspicious of its provincial officials for their stand during its wars. It was no longer able to exert the leadership it had previously commanded, but it was willing to change. The promulgation of the company law in 1904 as part of the Qing government's reform marks an institutional change which was as significant as the abolition of the imperial examinations. It recognized company incorporation and the principle of limited liability. That is to say, it limited state interference into business management and made raising capital easier. In another decade, with the collapse of the political order after the 1911 revolution, decentralization of power was complete, and industrial projects were no longer the privilege of a handful of senior officials.[13] Hence, with more private entrepreneurs investing in modern technology in the form of the joint-stock company, the pace of China's industrialization quickened, as historians have often noted.

There is no denial that China acquired steam technology with great difficulty. Yet, China's greatest difficulty was neither an inept government nor a conservative society, but the timing of technological change. By the 1895, the fast advancing Western technology was developed in research institutes and was sustained by science and mathematics. It took the collapse of the Qing empire for the new intuitional setting to be put in place for technology to take root.

Notes

1. *North China Herald*, 9th June 1893.
2. Xin Yuanou, Zhongguo jindai chuanbo gongye shi, pp. 205–208.
3. Sun Yutang (ed.), *Zhongguo jindai gongyeshi ziliao diyi ji*, pp. 1175, 1188.
4. Gao Shiliang (ed.) *Zhongguo jindai jiaoyu shi zhiliao: yangwu yundong shiqi jiaoyu* (Shanghai: Shanghai chubanshe, 1992), pp. 259–230.
5. For more details about the Chinese educational mission to the United States and Zhan Tianyou's life as a student in the U.S., see Edward J. M. Rhoads, *Stepping Forth into the World: The Chinese Educational Mission to the United States, 1872–81* (Hong Kong: Hong Kong University Press, 2011).
6. The English engineer was Claude William Kinder. He built China's second railway, the line between Tangshan and Xugezhuang in 1881.
7. For more details, see Ling Hongxun and Gao Zonglu (eds.), *Zhan Tianyou yu Zhongguo tielu* (Taipei: Institute of Modern History, Academia Sinica, 1977).
8. Takehiko Hashimoto, "Introducing a French Technological System: The Origin and Early History of the Yokosuka Dockyard," *East Asian Science, Technology and Medicien* 16 (1999), pp. 53–72.
9. For more details, see W. H. Brock, "The Japanese Connexion: Engineering in Tokyo, London and Glasgow at the End of the Nineteenth Century," *British Journal for the History of Science* 14:48 (1981), pp. 227–243;

W.H. Brock, Alex D. D. Craik, "Science and Technology in Nineteenth Century Japan: The Scottish Connection," *Fluid Dynamics and Research* 39 (2007), pp. 24–48.
10. Colin Latimer, "Kelvin and the Development of Science in Meiji Japan," in Raymond Flood, Mark McCartney and Andrew Whitaker (eds.), *Kelvin: Life, Labours and Legacy* (Oxford: Oxford University Press, 2008), pp. 212–223.
11. Xin Yuanou, *Zhongguo jindai chuanbo gongye shi*, pp. 176–186.
12. Sun Yutang, *Zhongguo jindai gongyeshi ziliao diyiji*, vol. 1, pp. 20–35.
13. For a thorough discussion on this issue, see David Faure, *China and Capitalism: A History of Business Enterprise in Moderna China* (Hong Kong: Hong Kong University Press, 2006), pp. 45–64.

References

Brock, W. H., "The Japanese Connexion: Engineering in Tokyo, London and Glasgow at the End of the Nineteenth Century," *British Journal for the History of Science* 14:48 (1981), pp. 227–243.
Craik, Alex D. D., "Science and Technology in Nineteenth Century Japan: The Scottish Connection," *Fluid Dynamics and Research* 39 (2007), pp. 24–48.
David Faure, *China and Capitalism: A History of Business Enterprise in Modern China* (Hong Kong: Hong Kong University Press, 2006).
Flood, Raymond, McCartney, Mark, and Whitaker, Andrew, (eds.), *Kelvin: Life, Labours and Legacy* (Oxford: Oxford University Press, 2008).
Gao Shiliang 高时良 (ed.) *Zhongguo jindai jiaoyu shi ziliao: yangwu yundong shiqi jiaoyu* 中国近代教育史资料: 洋务运动时期教育 (Shanghai: Shanghai chubanshe, 1992).
Gao Zonglu 高宗鲁 (eds.), *Zhan Tianyou yu Zhongguo tielu* 詹天佑與中國鐵路 (Taipei: Institute of Modern History, Academia Sinica, 1977).
Hashimoto, Takehiko, "Introducing a French Technological System: The Origin and Early History of the Yokosuka Dockyard," *East Asian Science, Technology and Medicien* 16 (1999), pp. 53–72.
North China Herald.
Sun Yutang 孫毓棠, *Zhongguo jindai gongyehsi ziliao diyi ji: 1840–1895* 中國近代工業史資料第一輯, 2 volumes (Beijing: Kexue chubanshe, 1957).
Xin Yuanou 辛元欧, *Zhongguo jindai chuanbo gongye shi* 中国近代船舶工业史 (Shanghai: Shanghai guji chubanshe, 1999).

GLOSSARY

Anhui 安徽
Annan yachuan 安南軋船
Anqing 安慶
Baoting 寶廷
Beiyang 北洋
Bi Ligan 畢利幹
Chang Qing 長慶
Chen Zhao'ao 陳兆翱
Chonghou 崇厚
chuanjian paoli 船堅砲利
Chuyang xuesheng 出洋學生
Cixi (Empress Dowager) 慈禧
cunliu 存留
Dagu 大沽
Daoguang 道光
Daye 大冶
Dengzhou 登州
Ding Gongchen 丁拱辰
Ding Richang 丁日昌
Ding Shoucun 丁守存
Dingyuan 定遠
diqi 地氣
dizhi 地支

Duban 督辦
Fanyi guan 翻譯館
fengshui 風水
fengyu zhen 風雨鍼
Fuzhou 福州
ganfa 感發
ganying 感應
Gansu 甘肅
Gezhi shushi 格致書室
Gezhi shuyuan 格致書院
Gezhi shuyuan keyi 格致書院課藝
Gewuxue 格物學
gong 功
Gong Yutang 龔毓棠
Gong Zhenlin 龔振麟
Gongcheng chu 工程處
Gonghang (*Cohong*) 公行
gonglü 功率
Gongyi xuetang 工藝學堂
Guandu shangban 官督商辦
Guangbing 廣丙
Guangding 廣丁 (*Fujing* 福靖)
Guangfangyan guan 廣方言館
Guangjia 廣甲
Guangyi 廣乙
Guangzhou 廣州
Guangxuehui 廣學會
Guo Songtao 郭嵩燾
Haijun yamen 海軍衙門
Hankou 漢口
Hanlin 翰林
hanshu biao 寒暑表
hanshu zhen 寒暑鍼
He Xin (Benjamin Hobson) 合信
Hou xuetang 後學堂
Hua Hengfang 華蘅芳
Hua Beiyu 華備鈺
Huaguan 畫館
Huai **Army** 淮軍

Huanghu 黃鵠
Huaxue 化學
Hubei 湖北
Huishi yuan 繪事院
Hunan 湖南
huo lunchuan 火輪船
Jiangsu 江蘇
Jiangxi 江西
jiansheng 監生
jicha chu 稽查處
Jilong 基隆
Jingxiang 京餉
Jinshi 進士
Jinxian dusi 儘先都司
Jinxian xuanyong xiancheng 儘先選用縣丞
Jingyuan 靖遠
Kaiji 開濟
Kaiping 開平
Kangxi 康熙
kaogong suo 考工所
kongqi 空氣
leng 冷
Li Hongzhang 李鴻章
Li Shanlan 李善蘭
Li Zhaotang 黎兆棠
li 力
li 里
li 理
li 禮
lijin 厘金
Lin Zexu 林則徐
Liu Kunyi 劉坤一
Liuhe congtan 六合叢談
Longwei 龍威 (*Pingyuan*平遠)
Luan Xueqian 欒學謙
mali 馬力
Mawei 馬尾
Min ju 閩局
Mohai shuguan 墨海書館

Nanyang 南洋
Nanchen 南琛
Nanrui 南瑞
Nian 捻
Ningbo 寧波
Niu Jian 牛鑑
Pan Shicheng 潘仕成
Pan Shirong (Pan Wenhai) 潘世榮 (潘文海)
Prince Gong 恭親王
qi 氣
qi 汽
Qian xuetang 前學堂
qianli chuan 千里船
qing liuzhi 輕流質
Qiying 耆英
re 熱
reliang 熱量
reqi 熱氣
reqi fasan 熱氣發散
reyuan 熱元
River Min 閩江
Rong Hong (Yung Wing) 容閎
rongrelü 容熱率 (heat capacity)
Shaanxi 陝西
Shen Baozhen 沈葆楨
shengshi zisheng rending yongbu jiafu 盛世滋生人丁永不加賦
shengyuan 生員
shui lunchuan 水輪船
Song Jin 宋晉
Song Yinxing 宋應星
suoli 縮力
Suzhou 蘇州
Tang Tingshu 唐廷樞
Tianjin 天津
tiangan 天干
tianqi 天氣
Tongwen guan 同文館
Tongzhi 同治
Wang Fucai 王輔才

Wang Tao 王韜
Wang Zhongyang 汪仲洋
Wannianqing 萬年清
Wei Han 魏瀚
Wei Yuan 魏源
weiyuan 委員
Wen'an chu 文案處
xixue 西學
wu dong zhi li 物動之力
wu xing 五行
Wu Zancheng 吳贊誠
wuxing wuzhi 無形無質
Xianfeng 咸豐
Xiang **Army** 湘軍
xiexiang 協餉
Xinjiang 新疆
Xu Jianyin 徐建寅
Xu Shou 徐壽
Xu Jianyin 徐建寅
xuanqi 玄氣
xuesheng 學生
Yang Yuhui 楊毓煇
Yangjiandu 洋監督
Yangqiang dui 洋槍隊
yangwu (*yiwu*) 洋務 (夷務)
Yangwu yundong 洋務運動
Yi Changhua 易長華
yinre 隱熱
yin-yang 陰陽
yinyang jiaogan 陰陽交感
Yijing 易經
Yipu 藝圃
yi 夷
yundong 運動
zashui 雜稅
Zeng Guofan 曾國藩
Zhan Tianyou 詹天佑
zhangli 漲力
Zhang Zhidong 張之洞

Zhejiang 浙江
Zheng Fuguang 鄭復光
Zhengyuan 鎮遠
zhengshui 正稅
zhiyong 致用
Zhiyuan 致遠
Zhili 直隸
zhiying chu 支應處
Zhongxi wenjain lu 中西聞見錄
zhongxue 重學
Zhoushan 舟山
zhuandong ruqi jiju 轉動入氣機具
zongjian'gong 總監工
zongban 總辦
Zongli yamen (*Zongli geguo tongshang shiwu yamen*) 總理衙門 (總理各國通商事務衙門)
Zuo Zongtang 左宗棠

Index[1]

A
Arnott, Neil, 50n44, 59–62, 64

B
Beijing, 5, 7, 15n30, 17n44, 34, 40, 47n4, 47n5, 47n6, 52n56, 73, 74, 82, 84n4, 98, 103, 105, 112, 126n29, 150, 197n21, 199n37, 211, 215
Beiying, 161, 162, 172n97, 173n110, 174n135, 187, 190, 203n85, 203n86, 204n98
Bingchuan qiji, 120–122
Bowu xinbian, 38, 39, 41, 50n44, 51n46, 52n59, 57, 59–67, 70, 72, 75–78, 81, 82, 84n8, 89n53, 95, 97, 111
Brown, Thomas, 100, 118, 119, 125n20

C
Calculus, 3, 4, 62, 87n21, 99, 101, 102, 119, 149–151, 212
Caloric, 60, 61, 64, 71, 80, 81, 85n14, 86n16, 88n36, 90n60, 101, 122
Calorimeter (*liangre qi*), 76, 100, 101, 109, 110, 114, 115, 122
Carnot, Nicolas Léonard Sadi, 62, 71, 72, 82, 86n20, 87n21
Chen Zhao'ao, 155, 164
Cixi, 6, 194, 204n92
Clausius, Rudolf, 71, 72, 82, 103
Cold (*leng*), 58, 61, 62, 64, 65, 67, 68, 72, 74, 75, 77, 78, 89n55, 96, 97, 111, 114
Compound engine, 62, 152–155, 157, 159, 161, 163, 170n71, 170n77, 171n82
Cunliu, 183

---

[1] Note: Page numbers followed by 'n' refer to notes.

D

D'Aiguebelle, Paul, 136–138, 142, 165n6, 167n18, 167n24, 176
Ding Gongchen, 30, 31, 33, 34, 36, 37, 39, 40, 49n27, 57, 120
Ding Richang, 170n78, 171n89, 191, 196n6, 197n14, 197n21
Drawing Office, the Fuzhou Navy Yard *(Huishi yuan)*, 144, 145, 150, 160

E

Edkins, Joseph, 106–108, 127n45, 127n46, 127n47
Energy, x, 3, 4, 46, 58, 59, 71, 72, 75, 76, 78, 80, 82, 97, 103, 106–110, 118–121, 210

F

Fryer, John, 41, 51n52, 52n56, 60, 85n11, 98–100, 102, 104–106, 108–113, 117, 119–122, 125n15, 125n21, 128n49, 129n69
Fuzhou Naval School *(Chuanzheng xuetang)*
  The School of Apprentice *(Yipu)*, 144, 149, 150, 162
  The School of Design *(Huishi yuan)*, 149, 150, 162
  The School of Naval Construction *(Qian xuetang)*, 104, 149–151, 154
  The School of Navigation *(Hou xuetang)*, 120, 149
Fuzhou Navy Yard *(chuanzheng ju)*, x, 45, 104, 120, 121, 123, 135–146, 175–195, 210–215

G

Gewu rumen, 73–76, 81–83, 89n48, 89n53, 89n55, 90n56, 90n57, 90n58, 95, 102, 103, 109, 111
Gezhi huibian, 105, 106, 111, 121, 122, 123n3
Gezhi qimeng, 106–110
Gezhi shuyuan, 111–113, 116, 117, 121, 128n55
Gezhi shuyuan keyi, 112, 129n64, 129n69
Giquel, Prosper, 136–138, 142–145, 150–152, 154–157, 161, 165n6, 165n8, 167n18, 167n24, 170n77, 171n93, 176, 180
Guangzhou, 4, 24, 25, 27, 29, 30, 33, 34, 36–38, 40, 46n3, 48n20, 49n34, 49n36, 59, 66, 95, 104, 105, 192, 209, 211, 212

H

Haijun yamen, 173n120
Hart, Robert, 189
Heat *(re)*, x, 11, 25, 31, 46, 57–83, 95–115, 149
Hobson, Benjamin (He Xin), 38, 39, 41, 50n43, 50n44, 59, 60, 62, 64–67, 71, 82
Hong Kong, 37–39, 43, 49n36, 52n63, 52n67, 59, 98, 105, 128n58, 143, 149, 215
Hua Hengfang, 40–44, 51n56, 70, 72, 82, 96, 98, 123n2, 129n70
Huanghu, 42, 43, 51n52, 57, 70, 72, 96, 97
Huaxue jianyuan, 100, 104, 125n15
Huo lunchuan (steamship), 24, 25, 28, 35

I
Ironclad, 154, 156, 170n76, 171n93, 186, 187, 189, 191, 192, 200n61, 201n63, 201n64, 201n65, 202n77, 202n79

J
Jiangnan Arsenal, 5, 6, 41, 45, 60, 97–102, 104–106, 111, 112, 118, 121, 122, 123–124n6, 125n21, 127n35, 127n45, 135, 142, 150, 151, 166n15, 167n19, 167n20, 167n24, 167n25, 167n29, 168n32, 168n34, 169n50, 169n51, 169n52, 169n53, 169n54, 169n57, 169n58, 169n59, 169n60, 169n64, 169n65, 170n72, 170n78, 177, 195, 215

K
Kaiji, 157, 174n135, 189, 190

L
Li (force), x, 75, 82, 83, 108, 109, 119, 120
Li (principle), 76, 96
Li Hongzhang, 5–9, 15n28, 43–45, 52n61, 52n63, 98, 103, 111, 113, 116, 128n58, 129n69, 151, 152, 155, 156, 158–159, 161, 162, 170n78, 172n97, 173n110, 177, 178, 180–182, 188–191, 193, 196n11, 197n14, 197n15, 198n30, 198n31, 198n33, 199n43, 201n62, 201n65, 202n74, 203n80, 203n85
Li Shanlan, 99, 107, 124n12
Lijin, 6, 184

M
Main, Thomas J., 100, 118, 119, 125n17, 125n20
Martin, W.A.P., 73–76, 78–83, 88n44, 89n48

N
Nanyang Fleet, 8, 120, 187, 189

P
Pingyuan Fleet, 159, 163, 190, 191
Prince Gong, 5, 6, 103
Prince Qing, x, xi, 1, 4–7, 9, 11, 12, 13–14n11, 24, 26, 29, 37, 40, 41, 43, 45, 46n2, 47n5, 48n20, 49n34, 60, 73, 82, 95–115, 138, 150, 152, 161, 163, 164, 181–184, 187–189, 191–195, 198n25, 199n47, 199n48, 200n53, 202n79, 210–212, 214–216

Q
Qi (steam), 31, 32, 39, 50n44, 58–70, 72, 80–83, 84n4, 84n5, 96–97
Qiji faren, 100, 102, 104, 121, 122, 125n19, 125n20

R
Rankin, William, 72, 103, 118, 119, 213
Rendel gunboat, 189
Rexue tushuo, 109, 110, 128n49

S
School for the Diffusion of Languages (*Guangfangyan guan*), 6, 106, 150, 151

School of Combined Learning (*Tongwen guan*), 6, 74, 75, 81, 82, 98, 103, 150
Sennett, Richard, 118, 119
Shanghai, 6, 28, 38, 43, 45, 50n44, 59, 82, 84n4, 89n44, 95, 98, 99, 103, 104, 106–108, 111, 116, 117, 124n12, 126n27, 128n58, 143, 149–151, 177, 178, 197n14, 198n33, 201n65, 204n97, 209, 211, 214, 215
Shen Baozhen, 136–138, 150–152, 154–156, 160, 161, 166n17, 168n41, 168n48, 169n68, 170n78, 176–182, 185–188, 191, 193, 196n6, 196n7, 196n8, 202n68, 203n80, 203n82
Song Jin, 152, 185
Steam engine, ix, 2–4, 7, 11, 23–30, 34–36, 38–40, 43–46, 52n59, 57, 59, 61, 62, 70–72, 76, 82, 89n48, 95–97, 100–105, 109, 111, 116, 121–123, 136, 139, 145, 150, 151, 160, 170n76, 171n82, 186, 199n45, 209, 210, 212
Stewart, Balfour, 106–108

T
Taiping Rebellion, 1, 5, 6, 40, 97, 135, 183, 184, 200n51
Taiwan Crisis (1874), 8, 9, 180, 186, 189, 193, 202n79, 203n80
Thermodynamics (*relixue*), ix, 3, 46, 70, 72, 73, 87n22, 95, 103, 118, 119
Thermometer (*hanshu zhen* or *hanshu biao*), 51n46, 58, 60, 61, 63, 64, 68, 70–72, 76, 78, 81, 82, 87n36, 90n58, 96, 97, 100, 107, 114–116, 122

Thomson, William (Lord Kelvin), 71, 72, 103, 106, 213
Tianjin, 7, 38, 40, 95, 105, 112, 171n93, 173n110, 178, 180, 212
Tidiao, 143
Translation Department, Jiangnan Arsenal (*Fanyi guan*), 6, 41, 60, 98, 104, 150, 151

W
Wannian Qing, 145, 179, 198n33
Watt, James, 2, 24, 39, 61, 62, 72, 76, 86n17, 86n18, 87n21, 100
Wei Han, 155, 164
Wei Yuan, 37, 46n2
Wu xing, 58–59, 70, 87n35
Wylie, Alexander, 98–102, 107, 111, 119, 124n12, 125n20

X
Xia'er guanzhen, 39, 67–72, 76, 78, 81, 82
Xiexiang, 183
Xu Jianyin, 102, 111
Xu Shou, 17n44, 40–44, 70, 72, 82, 96, 98–102, 104, 111, 112, 119, 123n4, 125n15, 125n20

Y
Yin-yang, 58–59, 65, 67, 68, 72, 73, 84n5, 116, 122

Z
Zeng Guofan, 5, 6, 40–45, 51n56, 52n63, 97, 98, 151, 169n64, 174n129, 176–178, 196n7, 197n21

Zhang Zhidong, 9, 173n111, 192, 211
Zheng Fuguang, 34, 35, 37, 39, 57
Zhengshui, 183
Zuo Zongtang, 5, 6, 135–138, 152, 157, 165n6, 165n8, 166n17, 167n20, 172n97, 175–178, 180, 181, 184, 185, 189–191, 193, 196n7, 197n17, 197n18, 203n85, 204n97

GPSR Compliance

The European Union's (EU) General Product Safety Regulation (GPSR) is a set of rules that requires consumer products to be safe and our obligations to ensure this.

If you have any concerns about our products, you can contact us on

ProductSafety@springernature.com

In case Publisher is established outside the EU, the EU authorized representative is:

Springer Nature Customer Service Center GmbH
Europaplatz 3
69115 Heidelberg, Germany

www.ingramcontent.com/pod-product-compliance
Lightning Source LLC
LaVergne TN
LVHW011006250326
834688LV00004B/100